SpringerBriefs in Computer Science

SpringerBriefs present concise summaries of cutting-edge research and practical applications across a wide spectrum of fields. Featuring compact volumes of 50 to 125 pages, the series covers a range of content from professional to academic.

Typical topics might include:

- A timely report of state-of-the art analytical techniques
- A bridge between new research results, as published in journal articles, and a contextual literature review
- A snapshot of a hot or emerging topic
- An in-depth case study or clinical example
- A presentation of core concepts that students must understand in order to make independent contributions

Briefs allow authors to present their ideas and readers to absorb them with minimal time investment. Briefs will be published as part of Springer's eBook collection, with millions of users worldwide. In addition, Briefs will be available for individual print and electronic purchase. Briefs are characterized by fast, global electronic dissemination, standard publishing contracts, easy-to-use manuscript preparation and formatting guidelines, and expedited production schedules. We aim for publication 8–12 weeks after acceptance. Both solicited and unsolicited manuscripts are considered for publication in this series.

**Indexing: This series is indexed in Scopus, Ei-Compendex, and zbMATH **

Keisuke Fujii

Machine Learning in Sports

Open Approach for Next Play Analytics

 Springer

Keisuke Fujii
Graduate School of Informatics
Nagoya University
Nagoya, Aichi, Japan

ISSN 2191-5768 ISSN 2191-5776 (electronic)
SpringerBriefs in Computer Science
ISBN 978-981-96-1444-8 ISBN 978-981-96-1445-5 (eBook)
https://doi.org/10.1007/978-981-96-1445-5

This work was supported by Japan Science and Technology Agency and Japan Society for the Promotion of Science.

This Springer imprint is published by the registered company Springer Nature Singapore Pte Ltd.
The registered company address is: 152 Beach Road, #21-01/04 Gateway East, Singapore 189721, Singapore

If disposing of this product, please recycle the paper.

To all those interested in learning-based sports analytics, may this book inspire you to join this exciting field, tackle its many challenges, and together, elevate the game

Foreword

When I studied computer science in the 1990s, the symbiosis of machine learning and sports was rather arcane and visionary if accepted at all. A friend of mine handed in a request to be eligible to complement his computer science major with a sports minor. The proposal must have been so uncommon and absurd at that time, that it was simply rejected by the Dean for that combination did not make sense to him at all. Looking back, the idea was good but the world was not ready yet.

Well, that was more than 20 years ago and the world moved on. In the meanwhile, academics like Keisuke raise the technical bar of machine learning approaches in sports level by level, sports scientists answer increasingly complex questions in data-driven fashions, and sports clubs and associations hire data scientists to tailor quantitative analyses to their own needs and philosophy. Still, the gap between academia and practice is still one to bridge but we are on the right path.

The most important thing about this book is that it may become a cornerstone of that very bridge. Keisuke offers a very complete overview of recent advances in sports analytics without being overly technical. He succeeds in presenting very complex mathematical relationships in an easily understandable way. You will hardly find any equation throughout the book and when you are interested in learning more about a particular approach or topic, Keisuke offers a great deal of carefully selected references containing all the tricks of the trade. This renders the book one of a kind and a nice read for everybody interested in sports analytics, irrespective of their technical skill set. If you had to settle on only a single book on sports analytics, this could be the one!

Sports scientists unfamiliar with contributions from machine learning will find all the resources and pointers they need to delve into their favorite topics. Likewise, computer scientists without a background in sports will certainly identify technical challenges and novel problem settings while all the folks in between may just enjoy the overview.

Whether you come from sports, computer science, or some other field, bringing together work and passion means happiness in life. What else could we ask for?

Enjoy the ride!

September 2024   Ulf Brefeld
                                                                        Professor
                                                    Leuphana Universität Lüneburg
                                                            Lüneburg, Germany

Preface

The world of sports is undergoing a significant transformation, driven by advancements in technology and data analytics. This book explores the field of learning-based sports analytics, focusing not on machine learning techniques themselves for sports, but on the application of machine learning in sports to enhance performance, training, tactics, strategy, refereeing, and fan engagement. It represents the culmination of years of research and practical applications, aiming to bridge the gap between theoretical concepts and real-world implementations. The motivation behind this work arises not just from the growing complexity and competitiveness in sports, but also from our intellectual curiosity and passion for improving sports performance. Traditional methods of data analysis, while valuable, no longer satisfy these deeper pursuits. By incorporating computer vision, data science, machine learning, and agent modeling, this book introduces a comprehensive analytical framework designed to enhance the predictive power and strategic depth of sports analytics. Moreover, this book emphasizes an open approach, where the use of open data, open code, and transparent, research-based methods drive the advancements of machine learning in sports, as already seen in fields such as computer vision and natural language processing.

It is important to clarify that this book is not a detailed guide on machine learning algorithms specifically tailored for sports, nor is it a manual on conducting machine learning research within sports contexts. Instead, it offers a broader perspective on how machine learning and other advanced technologies can be applied to sports analytics. Readers with a background in machine learning will find the content approachable and insightful, but the material is also designed to be accessible to those less familiar with the complexities of these algorithms. The focus is on the practical applications and implications of these technologies in the field of sports, rather than on the technical details of the algorithms themselves.

The book is structured into five chapters. Chapter 1 introduces the fundamentals and scope of learning-based sports analytics, setting the stage for the detailed discussions to follow. It includes key concepts and terminology, an overview of the role of data in sports analytics, and the modeling techniques that are foundational to the field. Chapter 2 focuses on data acquisition using computer vision, detailing how vast amounts of visual data are captured and processed to generate useful data for meaningful insights. Chapter 3 examines predictive analysis and play evaluation using machine learning, describing how these models can be used to predict results and evaluate plays. Chapter 4 explores the potential of agent-based modeling for play evaluation such as using reinforcement learning, highlighting the integration of real-world data with virtual simulations. Finally, Chapter 5 looks towards the future, discussing advanced research directions, practical deployment of learning-based analytics, and shaping the ecosystem of sports analytics.

Given the wide range of topics covered in this book, it was not feasible to include every relevant reference. Citations for representative works, particularly those centered around our group's research contributions, have been provided. However, readers interested in further exploration are encouraged to consult the cited literature. Chapter 5 synthesizes the discussion, addressing the challenges and opportunities when considering the integration of these diverse elements into a cohesive framework.

This book is intended for researchers, students, technologists, sports analysts, and anyone interested in the intersection of sports and technology. It is hoped that it serves as a valuable resource, inspiring further innovation and research in this exciting field.

October 2024

Keisuke Fujii
Associate Professor
Nagoya University
Nagoya, Japan

Acknowledgements

This book would not have been possible without the dedicated efforts of my coauthors across various papers and the insightful discussions we shared. I also wish to express my gratitude for the support provided by the JSPS KAKENHI and JST PRESTO funds in Japan, which have been helpful in advancing research in sports science, computer vision, machine learning, and trustworthy AI.

Additionally, I am thankful for the valuable discussions with reviewers in this book and our papers, as well as domain experts from various sports teams and professionals from various companies, whose knowledge and perspectives have significantly enriched this work. Finally, appreciation is expressed to the students of my research group, whose dedication and hard work have been invaluable throughout the course of this project. Their contributions have greatly enriched the research and analysis presented in this book. I am especially thankful to my wife and son for their understanding and encouragement. Without them, this work would not have been possible.

Contents

Acronyms

AI	Artificial Intelligence
CNN	Convolutional Neural Network
COCO	Common Objects in Context
CPS	Cyber-Physical Systems
DLT	Direct Linear Transformation
DMD	Dynamic Mode Decomposition
DTW	Dynamic Time Warping
E2E	End-to-End
GAN	Generative Adversarial Network
GAR	Group Activity Recognition
GFootball	Google Research Football
GNSS	Global Navigation Satellite System
GPS	Global Positioning System
GS-HOTA	Game State Higher Order Tracking Accuracy
GSR	Game State Reconstruction
HOG	Histogram of Oriented Gradients
HPUS	Hypothetical Possession Utilization Score
HRNet	High-Resolution Network
IMU	Inertial Measurement Unit
IRL	Inverse Reinforcement Learning
ITE	Individual Treatment Effect
LDA	Linear Discriminant Analysis
LPS	Local Positioning Systems
MAMDP	Multi-agent Markov Decision Process
MARL	Multi-agent Reinforcement Learning
MDP	Markov Decision Process
MOT	Multiple Object Tracking
NBA	National Basketball Association
NMF	Non-Negative Matrix Factorization
OBSO	Off-Ball Scoring Opportunity
PCA	Principal Component Analysis

RANSAC	Random Sample Consensus
Re-ID	Re-identification
RL	Reinforcement Learning
RNN	Recurrent Neural Network
RPS	Ranked Probability Score
RTK-GNSS	Real-Time Kinematic GNSS
SORT	Simple Online and Realtime Tracking
SPC	Shooting Payoff Computation
SVM	Support Vector Machine
TAS	Temporal Action Segmentation
t-SNE	t-distributed Stochastic Neighbor Embedding
VAR	Video Assistant Referee
VR	Virtual Reality
xG	Expected Goals
XR	Extended Reality
xSOT	Expected Probability of Shot On Target
YOLO	You Only Look Once

Chapter 1
What is Learning-Based Sports Analytics?

Abstract This chapter introduces the foundational concepts and frameworks of learning-based sports analytics, offering a comprehensive overview of the field. It begins by exploring key concepts such as data analytics, data acquisition, and data visualization, which are essential for understanding how data drives insights in sports analytics. The chapter then focuses on the critical role of data in sports, discussing the various types of data utilized, the methods and techniques for data collection, and the challenges associated with data acquisition and management. Following this, the chapter outlines the general background of modeling techniques, including the machine learning algorithms that are central to learning-based sports analytics and their practical applications in sports contexts. Lastly, it provides guidance on accessing and contributing to research in the field, addressing the complexities of navigating publication venues and offering practical opinions on disseminating and discovering impactful research. The foundation laid in this chapter prepares the way for a deeper exploration of advanced modeling techniques and the practical applications of learning-based sports analytics in the following chapters.

Keywords Machine learning · Data acquisition · Predictive modeling · Reinforcement learning · Team sports

1.1 Introduction

The field of sports analytics has recently seen remarkable growth and transformation. With the advent of machine learning, the potential to analyze and extract useful information from vast amounts of sports data has been expected. This chapter aims to introduce the concept of learning-based sports analytics, providing a foundational understanding of its principles and applications.

Learning-based sports analytics refers to the application of machine learning techniques to analyze and extract useful information from sports data, providing insights that can enhance performance and decision-making in sports [4]. In particular, this book focuses on their movements such as the next plays using their location and pose data as illustrated in Fig. 1.1. Unlike traditional approaches, learning-based

K. Fujii, *Machine Learning in Sports*,
SpringerBriefs in Computer Science, https://doi.org/10.1007/978-981-96-1445-5_1

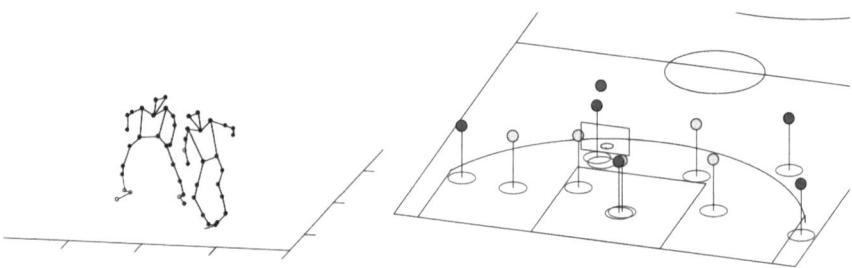

Fig. 1.1 Examples of pose and location data. (Left) An example of player pose data in a 1-vs-1 soccer dribble situation. The data was obtained from an optical motion capture system. (Right) An example of player location data in an NBA basketball game, which was obtained from a camera-based tracking system

ones leverage large and diverse datasets and complex algorithms to discover patterns and make predictions that are not immediately apparent. For example, player location and pose data for obtaining insights about the next plays will be challenging in traditional frameworks to mainly deal with outcomes and events. This field encompasses various technologies, including computer vision for data acquisition, predictive modeling to anticipate future events for play evaluation, and reinforcement learning for strategy optimization.

Regarding types of sports, we can roughly separate sports into individual sports and team sports, both of which require skillful, powerful, and/or accurate movements. When it comes to measuring movement data, some individual sports can be studied in controlled, laboratory environments. In contrast, many team sports require capturing the entire field of play, which presents a greater challenge. These properties can lead to issues when broadly analyzing various types of sports. It should be noted that some individual sports also take place over large areas during actual competitions. From the perspective of analyzing real-game scenarios where athletes perform at their best, focusing on team sports allows us to develop techniques that can be partially applied to individual sports as well.

This book basically focuses on team sports, as well as the complexities and dynamics involved in analyzing collective and skillful movements. In many team sports, such as soccer, basketball, handball, rugby, hockey, American football, volleyball, and baseball, data analysis faces several common challenges. Firstly, evaluation is often done through human observation or requires expensive measuring equipment, making it difficult for non-professionals to record large amounts of data. Even when vast amounts of data are available, predicting, explaining, and evaluating outcomes, as well as breaking down collective movements into individual contributions, can be challenging. This book focuses on invasion games like soccer and basketball, where data is accessible (as illustrated in Fig. 1.1) but analysis is particularly challenging due to the dynamic and collaborative nature of these sports.

1.1.1 Importance of Analytics in Modern Sports

In the current highly competitive sports environment, the role of analytics has become increasingly crucial. Teams and athletes are constantly seeking ways to obtain an edge over their competitors, and data-driven insights offer a significant advantage. Analytics can influence various aspects of sports, from player performance and injury prevention (e.g., [28]) to game strategy and fan engagement (e.g., [19, 20]). By utilizing advanced analytics, teams can make more informed decisions, optimize training programs (e.g., [29]), and develop strategies that are tailored to their strengths and the weaknesses of their opponents (e.g., [18]). Furthermore, analytics plays a vital role in enhancing the fan experience (e.g., [19]). Data-driven stories and insights can make the game more engaging for fans, providing them with a deeper understanding of the sport. Broadcasters and sports media leverage analytics to offer detailed analyses, enriching the narrative around games and events.

The use of analytics in sports is not a new concept. Historically, sports analytics began with basic statistical analyses, such as using player performance metrics and game statistics. These early efforts laid the groundwork for more sophisticated analyses. In the late 1970s, pioneers like Bill James [13] popularized the use of statistical methods in baseball through the development of Sabermetrics, which sought to objectively analyze player performance and game outcomes.

The revolution in sports analytics began with the development of digital technology and the availability of large datasets. In the 2000s, there was a significant interest in analytics across various sports, driven by the ability to collect and process vast amounts of data. The "Moneyball" era in baseball [14], exemplified by the Oakland Athletics' use of sabermetrics to build a competitive team on a limited budget, highlighted the potential of analytics to transform sports management.

A similar analytics-driven transformation occurred in basketball with the three-point revolution [16]. The Houston Rockets, under the guidance of general manager Daryl Morey, were among the pioneers of this approach, emphasizing three-point shooting and layups over mid-range two-point shots considering the points-per-shot ratio. This analytical approach has subsequently spread throughout the NBA, fundamentally changing the way the game is played [15] and highlighting the power of analytics to reshape sports strategies and outcomes.

As data collection methods became more advanced, so did the analytical techniques. The introduction of wearable technology, GPS (global positioning system) tracking, and video cameras allowed for the collection of detailed data on player movements, physiological metrics, and tactics in games. Such rich data necessitated more sophisticated analytical methods for the integration of machine learning and sports analytics.

1.1.2 Advancements in Machine Learning and Their Impact on Sports

Modeling sports behavior with machine learning surpasses traditional analysis by obtaining quantitative insights about human skill, team dynamics, and competitive strategies. Sports, with their structured and rule-based environments, present a rich field for machine learning applications. The challenges range from predicting player movements in basketball to evaluating and suggesting strategic actions in soccer, reflecting the comprehensive themes of this book: data acquisition with computer vision, play evaluation with machine learning prediction, and Potential play evaluation with learning-based agent modeling.

Although current expert insights in sports are mostly subjective and implicit because human skillful movements are sometimes hard to annotate by natural language, the goal is to quantify expert insights in sports, which signifies an essential development for the evolution of the field (Fig. 1.2). Capturing the excitement of motion in sports through the integration of real-world data and digital modeling [6], enhancing traditional sports data analysis. Historically, sports analytics focused on measuring and recognizing player positions and actions, such as shots or passes, within a physical space. However, experienced coaches and players go beyond mere observation; they predict future moves (e.g., [5]), evaluate the quality of motions (e.g., [25]), and suggest improvements (e.g., [17]). Currently, capturing these expert insights in digital form is challenging, but future advancements aim to make this knowledge universally accessible.

The digital transformation, converting the intricate knowledge and tactical wisdom of experienced professionals into digital formats, democratizes access to high-level sports strategies and practices, encouraging a learning and innovation ecosystem that is inclusive and expansive. The integration of expert insights with machine learning and data analytics tools amplifies the analytical depth and precision, encouraging idea exchange across diverse sports and related domains.

Furthermore, this digital infrastructure supports the development of predictive analytics and tailored training regimes, setting the stage for changes in athlete development, injury prevention, and performance enhancement (e.g., [1]). Ultimately, the digitization of expert sports knowledge catalyzes the formation of a well-informed, innovative, and interconnected global sports community, driving forward the fields of sports science and athlete development.

In the following Sect. 1.2, the key concepts of learning-based sports analytics are introduced such as data analytics, data acquisition, and data visualization. Section 1.3 focuses on the role of data in sports analytics, detailing the types of data used, the methods and techniques for data acquisition, and the challenges in data acquisition and management. Section 1.4 describes the general background of modeling techniques, including machine learning algorithms used in learning-based sports analytics and their applications to sports. Finally, Sect. 1.5 provides an overview of accessing and contributing to research in learning-based sports analytics, highlighting the com-

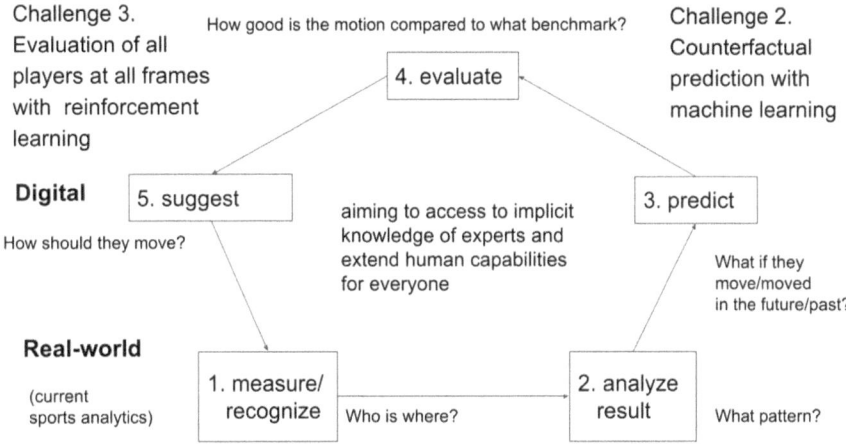

Challenge 3. Evaluation of all players at all frames with reinforcement learning

How good is the motion compared to what benchmark?

4. evaluate

Challenge 2. Counterfactual prediction with machine learning

Digital

5. suggest

How should they move?

aiming to access to implicit knowledge of experts and extend human capabilities for everyone

3. predict

What if they move/moved in the future/past?

Real-world

(current sports analytics)

1. measure/ recognize

Who is where?

2. analyze result

What pattern?

Challenge 1. Automatic data acquisition with computer vision

Fig. 1.2 Current and future perspective of sports analytics and their challenges. Traditionally, sports analytics involves (1) measuring and recognizing where players are on the field, then (2) analyzing what happens such as a shot or a pass in a physical space. The first challenge is the automation of the data acquisition process mainly with computer vision. In addition, experienced coaches and players can (3) predict future moves, (4) evaluate how well a motion was executed, and (5) suggest improvements. In learning-based sports analytics, counterfactual prediction with machine learning and evaluation and generation of all players at all frames such as with reinforcement learning are the second and third challenges, respectively. Ultimately, we aim to access to implicit knowledge of experts explicitly and extend human capabilities for everyone

plexities of navigating various publication venues and offering guidance on selecting appropriate platforms for disseminating and discovering research in the field.

1.2 Key Concepts and Terminology

In order to fully understand the scope and application of learning-based sports analytics, it is essential to grasp several foundational concepts. These concepts are interconnected and collectively contribute to the comprehensive framework of sports analytics. An example of data analytics flow is illustrated in Fig. 1.3. This section organizes these key concepts into structured categories to provide a clearer understanding of their roles and relationships as follows.

- **Data analytics** is the systematic computational analysis of data. In the context of sports, it involves examining and interpreting large volumes of data to discover patterns and insights. This is important for making informed decisions, optimizing performance, and gaining a competitive edge. Effective data analytics transforms raw data into actionable information, helping coaches, analysts, and players understand and improve their strategies and performances. To realize such data analytics,

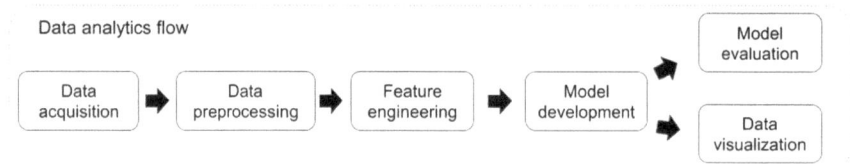

Fig. 1.3 An example of data analytics flow, including six key stages: data acquisition, data pre-processing, feature engineering, model development, model evaluation, and data visualization

we can consider data acquisition, data pre-processing, feature engineering, model selection, development, evaluation, and visualization as follows.

- **Data acquisition** refers to the process of collecting data from various sources. In sports analytics, this includes gathering information from video footage, wearable sensors, GPS tracking devices, and other monitoring technologies. Accurate and comprehensive data acquisition is essential for building a robust dataset that reflects the realities of sports performance. The quality of the collected data directly impacts the validity and reliability of subsequent analyses. The details are described in the next section, and in particular computer vision from video footage is discussed in the next chapter.

- **Data pre-processing** is the process of cleaning, transforming, and organizing raw data into a usable format. This step is important to ensure that the data used in the analysis is accurate, consistent, and free from errors or outliers. Data pre-processing involves tasks such as handling missing values, normalizing data, and converting data types. If we use the raw data initially, this process may take longer than expected. Proper pre-processing enhances the quality of the data, leading to more reliable and valid results in the subsequent analysis phase.

- **Feature engineering** involves using domain knowledge to extract meaningful features from raw data that is used for the analysis directly or enhance the performance of machine learning algorithms (the latter is related to learning-based analytics). In sports analytics, this could mean creating new variables that capture key aspects such as player performance, game conditions, or tactical formations. Effective feature engineering can not only interpret data effectively but also improve the predictive performance of models, making them more insightful and accurate.

- **Model development** involves choosing the appropriate machine learning and statistical models and building them to suit specific analytical tasks. This process includes evaluating different machine learning algorithms and training the models as necessary to determine the best fit for the data and the problem. Effective model selection and development are important for creating robust and accurate predictive models. Incorporating domain-specific knowledge into machine learning models ensures that the insights generated are relevant and applicable. This approach enhances the interpretability and accuracy of the models, making them more useful for coaches, players, and analysts.

- **Model evaluation** is the process of assessing the performance of models. Quantitatively, for example, this includes using various metrics to measure how well a model predicts outcomes compared to actual results. Qualitatively, it includes evaluating the interpretability, robustness, and relevance of the model in the specific context of its application. In sports analytics, model evaluation ensures that the developed models are reliable, interpretable, and can be trusted to provide accurate predictions and insights.

• **Data visualization** is the representation of data in graphical formats such as graphs, tables, maps, and charts. It helps in understanding trends, patterns, and insights from the data, making it easier to communicate findings to stakeholders. In sports analytics, data visualization can illustrate such as player movements, game statistics, and performance metrics, aiding in the decision-making process.

By connecting these key concepts with the data and modeling technologies in the following sections, a comprehensive framework for understanding learning-based sports analytics will be provided. In the next subsection, the critical role of data in sports analytics is described, focusing on the types of data used, the methods and technologies for data collection, and the challenges faced in data acquisition and management.

Strategy, Tactics, and Techniques in Sports

In the context of sports, strategies, tactics, and techniques are essential elements that explain the success of a team [11]. Strategies refer to comprehensive plans designed to achieve long-term objectives, such as changing lineups and formations in team sports. Tactics, on the other hand, are the specific actions or sequences of actions that teams employ during a game to execute their strategy. Examples include player positioning and play patterns in both offensive and defensive scenarios. Techniques are the fundamental skills required to execute these tactics effectively, including dribbling, passing, and shooting. Together, these three components form the backbone of competitive sports, with strategies providing the framework, tactics offering the means to adapt to the evolving dynamics of a game, and techniques ensuring the precise execution of these plans.

1.2.1 Positioning of This Book

This book offers a comprehensive exploration of learning-based sports analytics, distinguishing itself from recent reviews such as those focusing on methodology and evaluation in sports analytics [3], the potential of big data in soccer [10], and quantitative analysis in basketball [26, 30]. Unlike these works, which primarily concentrate on the application of analytics in specific sports contexts, this book focuses on the entire spectrum of sports analytics from the automatic acquisition of data to the application of advanced simulation techniques like reinforcement learning. Furthermore, it introduces my ideas on ecosystem development in sports analytics, emphasizing the creation of sustainable infrastructures that support open innovation and collaboration. By integrating these diverse elements, this book provides a unique and forward-looking perspective on the future of sports analytics, setting the stage for advancements in both practical deployment and theoretical research. Next, the fundamental roles that data plays are described, which are the important aspects and advanced applications of learning-based sports analytics.

1.3 The Role of Data in Sports Analytics

This section explores the critical role of data in sports analytics, focusing on the types of data used, the methods and technologies for data collection, and the challenges faced in data acquisition and management. Understanding these elements is essential for comprehending the complexities and potential of learning-based sports analytics.

1.3.1 Types of Data Used in Sports Analytics

Sports analytics relies on a diverse array of data types, each offering unique insights into athletic performance, strategy, and outcomes. Figure 1.4 gives examples of a soccer game scenario with players' locations, poses, and a "pass" event. The primary types of data are as follows.

- **Video data** recordings from games and practices provide rich visual information that can be analyzed to track player movements, actions, and interactions. In the context of sports, it is common practice for coaches, players, and analysts to review video footage to gain insights and improve performance. Computer vision techniques are often employed to extract meaningful insights from video data,

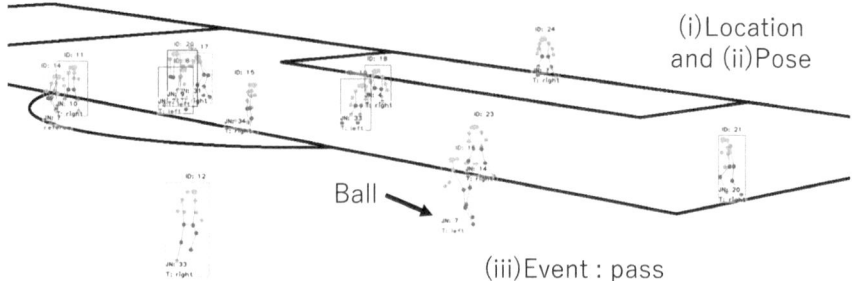

Fig. 1.4 Examples of data types. It illustrates a soccer game scenario where players' locations, events, and poses are identified using bounding boxes, a label, and keypoints. The event captured in this instance is a "pass". This is the original figure using the SoccerNet data [21]

automating the analysis process and discovering patterns that may not be immediately apparent to the human eye. The detailed usage of video footage will be introduced in the next chapter.

- **Match or game data** includes fundamental records of key actions and occurrences during a game, such as goals, assists, substitutions, and fouls, which are defined in the rules of each sport, typically recorded by humans. This data provides fundamental information about the game and forms the backbone of performance analysis, helping teams optimize their tactics and improve overall performance. Match data is usually open and can be scraped from the web.

- **Event data** is the detailed records of specific actions or events and occurrences during a game, such as passes and intercepts (their names and definitions depend on sports and organizations) in addition to match data, typically obtained by humans. This data includes precise timestamps and (often) spatial coordinates, allowing us to understand the context and impact of each action within the flow of the game. For example, event data captures the origin and destination of each pass, the location, (sometimes) type, and outcome of each shot, as well as details about defensive actions like tackles and interceptions. It also records fouls, including the players involved and any disciplinary actions, as well as goal-scoring events and the circumstances leading to the goals. Event data is crucial for understanding player performance and team strategy. By combining event data with tracking data, analysts gain a comprehensive view of the game's dynamics, leading to more informed and strategic decision-making in sports analytics. Some data provider companies release certain event data. For example, WyScout and StatsBomb have made available data for soccer events.

- **Location or tracking data** includes the spatial coordinates of players and the ball, capturing their locations and movements on the field or court using devices in the following subsection. It is essential for analyzing tactics and player positioning. If location data is collected at regular intervals, and if sampled at sufficiently short and approximately equal intervals, it is called **tracking data** and can also be used to compute movement speed and acceleration. This provides deeper insights into player dynamics and performance. For example, Metrica provides tracking data in three soccer games,[1] and SportVU tracking data is available in 631 basketball games.[2]

- **Pose data** in sports analysis refers to the detailed information about the body positions and movements of athletes, typically captured by such as motion capture, computer vision, or wearable sensors. This data includes the coordinates of key body joints (e.g., knees, elbows, hips) at various time points, allowing for an in-depth analysis of an athlete's posture, technique, and biomechanics. Pose data is essential for understanding how athletes move, identifying potential areas for improvement, and preventing injuries by detecting abnormal or suboptimal movements. It can be used to enhance training programs, optimize performance, and tailor coaching strategies to individual needs. By providing a precise and dynamic

---

[1] https://github.com/metrica-sports/sample-data/.

[2] https://github.com/rajshah4/BasketballData.

representation of an athlete's movements, pose data enables a more granular and actionable analysis, contributing to better overall performance and reduced injury risks in sports. Although various applications are considered, pose data cannot be annotated at a game-level, making its use much more limited.

- **Other sensor data** in sports analysis includes a range of metrics collected from devices such as heart rate monitors, accelerometers, and lactate sensors, which may be sometimes wearable or integrated into equipment. This comprehensive data is vital for real-time monitoring of an athlete's physical condition, identifying fatigue, and preventing overtraining or injuries. Sensor data offers continuous and detailed streams that can be analyzed to tailor training programs, optimize performance, and ensure athlete well-being. However, sensor data is usually unavailable in public, except for GPS data.

1.3.2 Data Collection Methods and Technologies

Effective data collection is fundamental to sports analytics. Here, the methods and technologies used to collect data are introduced.

Markerless Tracking Systems

Advanced computer vision technologies enable the capture of motion data without the need for any markers or sensors, making them particularly useful for recording natural movements during actual gameplay. Historically, motion capture involved manual digitizing (e.g., until about 2000s [7]), where movements were tracked frame-by-frame by hand, or the use of reflective markers in optical motion capture systems (currently used as most accurate methods in sports sciences). These methods, while accurate, were time-consuming, costly, and often impeded natural movement. In comparison, markerless systems offer significant advantages: they allow for natural, unrestricted movement, reduce setup time, and lower costs. Notable examples to track player and ball location data in professional team sports include SportVU in basketball (e.g., used in [12]), and TRACAB in soccer (e.g., used in [27]). In the computer vision area, multi-object tracking technologies have been developing. Pose estimation methods from video footage without markers have also advanced the field, with notable examples including OpenPose [2]. These computer vision techniques are discussed in Chap. 2. Additionally, technologies such as LiDAR and depth sensors enhance the capability of capturing precise motion data, potentially improving the analysis of player movements and team tactics with less occlusion.

Global and Local Positioning Systems

These technologies track the location and movement of players in real-time, providing positional data. GPS is commonly used in outdoor sports. In contrast, local positioning systems (LPS) can be used in indoor settings and more accurate but relatively be more expensive than GPS. Recent advancements in global navigation satellite systems (GNSS), including real-time kinematic GNSS (RTK-GNSS), have significantly improved the accuracy of positional data. GNSS refers to satellite systems that provide global positioning services, with GPS being one of the most well-known examples. RTK-GNSS enhances GNSS by using a fixed base station to provide real-time corrections to the positional data received from satellites. This technology can achieve centimeter-level accuracy, making it highly suitable for applications that require precise tracking and positioning. Consequently, RTK-GNSS can offer a more accurate alternative to traditional GPS, particularly in outdoor sports settings where high precision is essential.

Other Sensors

Other devices such as accelerometers, gyroscopes, and heart rate monitors are worn by athletes to measure various performance metrics. These sensors provide real-time data that can be used for immediate feedback and long-term analysis. Heart rate monitors track cardiovascular performance, providing insights into an athlete's fitness level, endurance, and recovery times. Accelerometers measure the acceleration of the body, capturing data on speed, agility, and the intensity of movements. Gyroscopes, which detect rotational movements, help in analyzing balance and coordination. In addition, magnetometers, which measure the Earth's magnetic field, are sometimes used to detect orientation relative to the Earth's magnetic poles. When combined with accelerometers and gyroscopes, these sensors form an Inertial Measurement Unit (IMU), commonly used to provide comprehensive motion analysis. Lactate sensors measure the concentration of lactate in the blood, offering crucial information about an athlete's anaerobic threshold and overall metabolic condition.

1.3.3 Challenges in Data Acquisition and Management

Despite the advances in data collection technologies, several challenges remain. For example:

- **Data Quality**: Ensuring the accuracy and reliability of collected data is crucial. Automatically captured data, such as that from wearable sensors and cameras, often contains noise and errors compared to ground truth data created using optical motion capture systems or manual corrections. This automatic data acquisition introduces inaccuracies that necessitate careful pre-processing to filter out errors and improve data quality.

- **Data Volume**: The volume of data generated during sports events can significantly impact the choice of machine learning models used for analysis. Large datasets enable the use of complex models, such as deep learning, which require extensive data to train effectively. Conversely, smaller datasets might necessitate the use of simpler models, such as linear regression or decision trees, which are less data-intensive. Therefore, the amount of data available directly influences the modeling approach and the potential insights that can be derived from the analysis. Unfortunately, in cases where data is not provided by professional sports data providers, the volume of available data is often limited.

- **Privacy Concerns**: Collecting and analyzing personal data, especially visual and pose information, can identify individuals, raising significant privacy issues. Professional league data provided by official data providers is typically covered under contracts with the leagues. However, when researchers collect data themselves, such as in university studies, ethical approval from an institutional review board is required. Additionally, separate consent is necessary for the public release of any collected data. Ensuring that data is collected and used ethically, with proper consent and safeguards, is essential.

- **Integration of Data Sources**: Combining data from multiple sources (e.g., wearable sensors, video, and GPS) to create a comprehensive dataset can be complex. This integration is crucial for generating holistic insights but requires various data fusion techniques such as sensor fusion, data alignment, synchronization, interpolation, and format integration. Such (pre-) processing steps can be time-consuming, adding to the complexity of the main task.

1.3.4 Addressing the Complexities of Sports Movement Analysis

Modeling sports behavior with machine learning has the potential to surpass traditional analysis by exploring the complexities of human skill, team dynamics, and competitive strategies. For example, capturing the intricate movements in a basketball 1-vs-1 situation (e.g., [8]) or the coordination of a soccer team (e.g., [25]) requires advanced technologies and methodologies. This complexity highlights the importance of accurate and comprehensive data collection.

Technologies such as wearable sensors and markerless tracking systems play a vital role in capturing these complex movements. Despite these advancements, challenges such as data noise, high costs, and the practical difficulties of attaching sensors to athletes persist. Additionally, ensuring that data accurately reflects the dynamic and interactive nature of sports remains a critical task.

In summary, understanding the types of data, the methods for collecting it, and the challenges involved are essential for advancing the field of learning-based sports analytics. In the next section, various modeling techniques used in learning-based sports analytics will be explored, such as how these methods transform raw data into actionable insights, enhance player performance, and inform strategic decisions.

1.4 Modeling Techniques for Learning-Based Sports Analytics

In this section, various modeling techniques are introduced, which are used in the analysis of learning-based sports analytics. Generally, modeling refers to the abstraction and representation of real-world phenomena in a way that makes them easier to understand. Our real-world includes numerous complex systems and processes, such as climate change, market trends, and human behavior, which are difficult to comprehend solely through intuition or experience. Modeling helps us understand, predict, and control these phenomena. Although detailed examples will be presented in Chap. 3, this section provides an overview of the essential modeling framework, which is needed to understand the techniques and applications discussed also in Chap. 5.

In the context of learning-based sports analytics, the concepts of forward and inverse problems play a crucial role in understanding and analyzing sports performances as illustrated in Fig. 1.5. A forward problem involves generating outcomes from known causes or models, such as predicting the results of specific tactics or formations in a soccer game. This approach, known as forward analysis, is often used in simulations where predefined strategies are tested to observe their potential impact. Conversely, an inverse problem involves deducing the underlying strategies or causes from observed outcomes, such as analyzing game data to discover the tactics that led to a particular result. This approach, referred to as inverse analysis, is commonly employed by coaches and analysts to understand actual team performance. Both forward and inverse analysis are essential in learning-based sports analytics to bridge the gap between simulating future outcomes and understanding the current tactics. Further details on these methods will be discussed in Chap. 4.

Fig. 1.5 Conceptual diagram of forward and inverse problems. Generating data from a model is a forward problem and is referred to as forward analysis, while estimating the model from data is an inverse problem and is referred to as inverse analysis

Models can be categorized into several types, including mathematical models, rule-based models, and machine learning models as follows. The detailed examples and their integrations will be presented in Chaps. 3 and 5.

1.4.1 Mathematical Models

Mathematical modeling is used across various fields, such as physics, economics, and biology, to represent real-world phenomena through mathematical expressions (equations or functions). These models are constructed based on physical laws and/or clear mathematical principles to predict future behaviors of systems and guide system design. In sports, for example, models based on the equations of motion can predict the trajectory of a ball or an athlete's movement during play (e.g., [22] in soccer). Another example is spatial partitioning methods, such as Voronoi diagrams (e.g., [24] in soccer), applied in team sports to analyze player optimal positioning and tactics. These and other examples are introduced in Chap. 3.

1.4.2 Rule-Based Models

A rule-based model is a modeling approach where the underlying "rules" governing a system or behavior are explicitly programmed into a computer by humans. By rule-based modeling, complex phenomena are broken down into simpler, more manageable sets of rules. For example, in sports analytics, a rule-based model can be used to classify and evaluate player movements based on predefined strategies or game rules (e.g., [9, 23] in basketball. For the details, see Chap. 3). The computer follows these programmed rules to simulate and analyze different scenarios, such as determining the best possible player positioning during a match (e.g., various simulators introduced in Chap. 4). This approach is particularly useful when the system being modeled operates under clear, well-defined guidelines that can be accurately captured through programming.

1.4.3 Machine Learning Models

Machine learning models automatically extract useful patterns or knowledge from input data, enabling predictions or classifications on unknown data. Unlike traditional programming, where features and rules are explicitly defined by humans, machine learning models "learn" these from data. This approach is widely applicable in areas such as image recognition, natural language processing, medical diagnostics, and autonomous driving, though it requires large amounts of data and computational resources. In the context of sports analytics, machine learning models can be divided into two major categories: extracting features from data and simulating or controlling specific actions [4]. Most feature extraction techniques are categorized into inverse analysis. Typical machine learning approaches include unsupervised learning, supervised learning, reinforcement learning, and their combinations.[3]

Regarding approaches to extracting features from data, unsupervised learning involves learning without using target variables (e.g., play types or scores in sports data analysis), which includes dimensionality reduction and clustering. Supervised learning involves learning to align with target variables. When the target variable is discrete (e.g., play types), it is called classification; when continuous (e.g., position data or scores), it is called regression. Like school tests, this method typically involves using a separate test dataset for validation to prevent overfitting (see Sect. 3.4).

Simulating and controlling actions can be done using pattern-based or agent-based approaches. Pattern-based methods use supervised learning to predict player actions and movement trajectories for simulations. In contrast, agent-based methods involve modeling the agents themselves, often applying reinforcement learning (RL, see also Chap. 4). RL models agents that learn to perform actions to maximize rewards, deriving policies from states. While RL concepts can also be applied to inverse analysis, it is generally difficult to model human players accurately, leading to applications in state or action evaluation, policy, or reward estimation (known as imitation learning or inverse reinforcement learning).

In summary, the modeling concepts and problem settings discussed in this chapter form the basis for understanding the simulation and analysis of human behavior in sports. These models, whether mathematical, rule-based, or machine learning-based, share the common goal of predicting and understanding unknown phenomena and data. By leveraging the strengths of each approach, we can select the most appropriate model to solve specific challenges in sports analytics, which will be discussed in Chaps. 3, 4, and 5. In particular, the integration of data collections (in Chap. 2) and various modeling (in Chaps. 3 and 4) for real-world applications are discussed in Chap. 5.

1.5 Accessing and Contributing to Learning-Based Sports Analytics Research

In the field of learning-based sports analytics, finding relevant research papers can be a complex and often confusing task when searching for conferences and journals. Unlike other scientific and engineering research domains, where top journals and major international conferences are well-established, the landscape of sports analytics lacks a clear consensus on top venues for publication. This lack of clarity makes it challenging for researchers, particularly those new to the field, to identify where to publish or where to find the most impactful research.

---

[3] It should be noted that these distinctions are often blurred in practice, as described in Chap. 3.

For those primarily interested in reading rather than writing papers, understanding the different types of publications in this field is crucial. Broadly speaking, publications can be categorized into six types:

1. Sports science journals
2. Journals regarding computer science and engineering
3. Journals covering both sports, computer, and engineering science
4. Conferences for sports science and analytics
5. Conferences in computer science and engineering
6. Sports workshops at conferences in computer science and engineering

It is important to note that most sports science conferences typically do not publish full papers; instead, they focus on abstract submission and information exchange, which can make them less suitable for finding detailed research papers. The main challenge lies in the scattered nature of these papers across different venues and the fact that the best sources of information depend on our specific research objectives.

Particularly for categories 1–3, it is not uncommon to question why certain papers were published in specific journals. This often has more to do with the preferences and constraints of the authors rather than the needs of the readers. While Google Scholar is a useful tool, arXiv can also be a valuable resource for finding preprints, especially for papers in categories 2, 3, 5, and 6. However, one must be cautious, as arXiv papers are not peer-reviewed, which means that while they offer early access to research, their quality is not guaranteed.

When seeking the latest developments in the field, international conferences tend to provide more up-to-date information. Among the recommended conferences and workshops are CVSports[4] at CVPR and MMSports[5] at ACM MM in the computer vision and multi-media domains, and MIT SSAC[6] and MLSA[7] at ECML-PKDD in the sports analytics domain. These conferences and workshops are well-known and are held regularly, providing a platform for the latest research.

For those considering entering this field or seeking to understand the considerations of researchers in sports analytics, it is essential to recognize that only a small number of researchers consistently publish in this area globally. The field is still developing, with many research challenges yet to be addressed, making it a promising area for future work. Unlike other scientific and engineering fields, where the publication process is more straightforward, researchers in sports analytics must navigate a complex landscape where seemingly similar studies might be submitted to entirely different venues based on subtle differences in focus. Given the challenges of the publishing process, choosing the right venue is crucial to avoid the frustration of a misplaced submission.

---

[4] International Workshop on Computer Vision in Sports, https://vap.aau.dk/cvsports/.

[5] Multimedia Content Analysis in Sports, a webpage in 2024 is: http://mmsports.multimedia-computing.de/mmsports2024/.

[6] MIT Sloan Sports Analytics Conference, https://www.sloansportsconference.com/.

[7] Workshop on Machine Learning and Data Mining for Sports Analytics, a webpage in 2024 is: https://dtai.cs.kuleuven.be/events/MLSA24/.

The experience of the author involved submitting to a wide range of venues, often through trial and error, and the author hopes to share these insights to help others avoid common pitfalls. The fundamental reason for publishing is to disseminate new scientific or engineering knowledge. To do this effectively, one must thoroughly understand prior research and conduct studies in a way that convinces peer reviewers of the validity of the findings. While the author believes that the quality of research is more important than the venue of publication, the reality is that community recognition plays a significant role in securing academic activities. The challenge in sports analytics and machine learning is that the most respected venues in this field (categories 4 and 6) are not always recognized in adjacent science and engineering fields. Therefore, when selecting a venue for submission, the content of the research, the context of the authors, and the broader goals of the research project should be carefully considered.

In summary, for researchers new to this field or those with a background in related areas, I hope this overview provides useful guidance on where to submit or find research in sports analytics. Despite the current complexity of the field, it remains an area rich with potential and opportunities for groundbreaking work.

1.6 Summary

This chapter has outlined a general introduction and the critical elements in learning-based sports analytics and the complexities involved in navigating the diverse publication landscape. By providing a structured approach to understanding the key concepts and challenges in data acquisition, processing, and analysis, the chapter equips researchers with the knowledge needed to make decisions about where to focus their efforts and how to contribute effectively to the field. The next chapter will focus on the role of computer vision in sports analytics, specifically focusing on how automatic data acquisition technologies are performed in sports.

References

1. Beal, R., Norman, T.J., Ramchurn, S.D.: Artificial intelligence for team sports: a survey. Knowl. Eng. Rev. **34**, e28 (2019)
2. Cao, Z., Hidalgo, G., Simon, T., Wei, S.-E., Sheikh, Y.: OpenPose: realtime multi-person 2D pose estimation using Part Affinity Fields (2018). arXiv:1812.08008
3. Davis, J., Bransen, L., Devos, L., Jaspers, A., Meert, W., Robberechts, P., Van Haaren, J., Van Roy, M.: Methodology and evaluation in sports analytics: challenges, approaches, and lessons learned. Mach. Learn. 1–34 (2024)
4. Fujii, K.: Data-driven analysis for understanding team sports behaviors. J. Robot. Mechatron. **33**(3), 505–514 (2021)
5. Fujii, K., Takeishi, N., Kawahara, Y., Takeda, K.: Decentralized policy learning with partial observation and mechanical constraints for multiperson modeling. Neural Netw. **171**, 40–52 (2024)

6. Fujii, K., Tsutsui, K., Scott, A., Nakahara, H., Takeishi, N., Kawahara, Y.: Adaptive action supervision in reinforcement learning from real-world multi-agent demonstrations. In: 16th International Conference on Agents and Artificial Intelligence (ICAART'24), vol. 2, pp. 27–39 (2024)

7. Fujii, K., Yamada, Y., Oda, S.: Skilled basketball players rotate their shoulders more during running while dribbling. Percept. Mot. Skills **110**(3), 983–994 (2010)

8. Fujii, K., Yamashita, D., Kimura, T., Isaka, T., Kouzaki, M.: Preparatory body state before reacting to an opponent: Short-term joint torque fluctuation in real-time competitive sports. PLoS ONE **10**(5), e0128571 (2015)

9. Fujii, K., Yokoyama, K., Koyama, T., Rikukawa, A., Yamada, H., Yamamoto, Y.: Resilient help to switch and overlap hierarchical subsystems in a small human group. Sci. Rep. **6** (2016)

10. Goes, F.R., Meerhoff, L.A., Bueno, M., Rodrigues, D.M., Moura, F.A., Brink, M.S., Elferink-Gemser, M.T., Knobbe, A.J., Cunha, S.A., Torres, R.S., et al.: Unlocking the potential of big data to support tactical performance analysis in professional soccer: A systematic review. Eur. J. Sport Sci. **21**(4), 481–496 (2021)

11. Gréhaigne, J.-F., Godbout, P., Bouthier, D.: The foundations of tactics and strategy in team sports. J. Teach. Phys. Educ. **18**(2), 159–174 (1999)

12. Hojo, M., Fujii, K., Inaba, Y., Motoyasu, Y., Kawahara, Y.: Automatically recognizing strategic cooperative behaviors in various situations of a team sport. PLoS ONE **13**(12), e0209247 (2018)

13. James, B.: The New Bill James Historical Baseball Abstract. Simon and Schuster (2010)

14. Lewis, M.: Moneyball: The Art of Winning an Unfair Game. WW Norton & Company (2004)

15. Mandić, R., Jakovljević, S., Erčulj, F., Štrumbelj, E.: Trends in NBA and euroleague basketball: analysis and comparison of statistical data from 2000 to 2017. PLoS ONE **14**(10), e0223524 (2019)

16. Mills, J.: Decision-Making in the NBA: The Interaction of Advanced Analytics and Traditional Evaluation Methods. Ph.D thesis, University of Oregon (2015)

17. Nakahara, H., Tsutsui, K., Takeda, K., Fujii, K.: Action valuation of on-and off-ball soccer players based on multi-agent deep reinforcement learning. IEEE Access **11**, 131237–131244 (2023)

18. Nikić, A., Topalović, A., Bach, M.P.: From data to decision: machine learning in football team management. In: 2024 47th MIPRO ICT and Electronics Convention (MIPRO), pp. 1059–1064. IEEE (2024)

19. Prins, M.: Improving content discovery and viewer engagement with ai. In: Proceedings of the 1st Mile-High Video Conference, pp. 132–132 (2022)

20. Sarkhoosh, M.H., Dorcheh, S.M.M., Gautam, S., Midoglu, C., Sabet, S.S., Halvorsen, P.: Soccer on social media (2023). arXiv:2310.12328

21. Somers, V., Joos, V., Cioppa, A., Giancola, S., Ghasemzadeh, S.A., Magera, F., Standaert, B., Mansourian, A.M., Zhou, X., Kasaei, S., et al.: Soccernet game state reconstruction: End-to-end athlete tracking and identification on a minimap. In: Proceedings of the IEEE/CVF Conference on Computer Vision and Pattern Recognition, pp. 3293–3305 (2024)

22. Spearman, W., Basye, A., Dick, G., Hotovy, R., Pop, P.: Physics-based modeling of pass probabilities in soccer. In: Proceeding of the 11th MIT Sloan Sports Analytics Conference (2017)

23. Supola, B., Hoch, T., Baca, A.: Modeling the offensive-defensive interaction and resulting outcomes in basketball. PLoS ONE **18**(2), e0281467 (2023)

24. Taki, T., Hasegawa, J., Fukumura, T.: Development of motion analysis system for quantitative evaluation of teamwork in soccer games. In: Proceedings of 3rd IEEE International Conference on Image Processing, vol. 3, pp. 815–818. IEEE (1996)

25. Teranishi, M., Tsutsui, K., Takeda, K., Fujii, K.: Evaluation of creating scoring opportunities for teammates in soccer via trajectory prediction. In: International Workshop on Machine Learning and Data Mining for Sports Analytics, pp. 53–73. Springer (2022)

26. Terner, Z., Franks, A.: Modeling player and team performance in basketball. Ann. Rev. Stat. Appl. **8**(1), 1–23 (2021)

27. Toda, K., Teranishi, M., Kushiro, K., Fujii, K.: Evaluation of soccer team defense based on prediction models of ball recovery and being attacked: a pilot study. PLoS ONE **17**(1), e0263051 (2022)
28. Eetvelde, H.V., Mendonça, L.D., Ley, C., Seil, R., Tischer, T.: Machine learning methods in sport injury prediction and prevention: a systematic review. J. Exp. Orthop. **8**, 1–15 (2021)
29. Wang, C., Du, C.: Optimization of physical education and training system based on machine learning and internet of things. Neural Comput. Appl. 1–16 (2022)
30. Zhou, Y., Li, T.: Quantitative analysis of professional basketball: a qualitative discussion. J. Sports Anal. **9**(4), 273–287 (2023)

Chapter 2
Computer Vision for Sports Analytics

Abstract Recent advancements in computer vision have significantly impacted sports analytics by automating the collection, analysis, and interpretation of data from sports video footage. Traditionally, data collection and labeling in sports has relied heavily on manual effort, which is both time-consuming and costly. However, computer vision offers a more efficient alternative by employing advanced algorithms to extract meaningful information from video footage, thus enabling detailed insights into player movements and team tactics. Computer vision is applied across various tasks including field registration, object tracking, action recognition and detection, and pose estimation. These tasks leverage machine learning models to handle large volumes of visual data. This chapter explores how these technologies are transforming sports analytics, introducing interesting research examples and highlighting the importance of automated data collection for sports analytics.

Keywords Field registration · Object tracking · Player identification · Action recognition · Pose estimation

2.1 Introduction

In recent years, computer vision has emerged as a powerful tool in the field of sports analytics, advancing the way data is collected, analyzed, and interpreted. Computer vision is a research field involving the use of algorithms and techniques to enable computers to interpret and understand visual information from the world. In sports, this means capturing and analyzing video footage to extract valuable insights about player movements and tactical elements. However, much of data acquisition, including raw data collection and labeling relies on manual effort, which is time-consuming and labor-intensive. While expensive equipment is sometimes used, financial constraints limit these use to only top-tier professionals. Therefore, there is a growing need for automated data acquisition using computer vision technologies.

The application of computer vision in sports analytics spans a wide range of tasks, including field registration, object tracking, action recognition, and pose estimation. These tasks leverage advanced machine learning models to process vast

21

K. Fujii, *Machine Learning in Sports*,
SpringerBriefs in Computer Science, https://doi.org/10.1007/978-981-96-1445-5_2

amounts of visual data, providing detailed and meaningful information that was previously unattainable with traditional methods. By automating the extraction of meaningful information from video footage, computer vision allows for more comprehensive insights into sports performance and strategy. International workshops such as CVSports and MMSports on computer vision in sports continually introduce the latest approaches for efficiently collecting and analyzing these data. This chapter explored how these technologies are transforming the future of sports analytics, showing interesting research examples. Although a comprehensive list of commercial devices and applications is provided in [166], the focus of this chapter is to introduce openly shared technologies rather than less transparent commercial devices. In the following, an overview of key computer vision tasks is provided. Then, Sect. 2.2 describes the details of essential computer vision elements for analyzing sports data. Section 2.3 explores advanced applications and integrations of computer vision techniques. Finally, Sect. 2.4 discusses future directions and potential advancements in computer vision for sports analytics.

2.1.1 Overview of Key Computer Vision Tasks for Sports Analysis

Following the introduction of computer vision's role in sports analytics, it is essential to consider fundamental tasks for sports analytics. The subsequent tasks form the backbone of computer vision applications in this domain. Here, field registration, tracking, re-identification, action recognition and detection, and pose estimation are briefly introduced. These tasks are crucial for generating event data, tracking data, and pose data, which are essential for comprehensive sports analytics. The detailed explanations are described in the next subsection. Figure 2.1 illustrates the application of field registration, tracking, and identification in a soccer match [150].

Field Registration involves the alignment of the captured video footage with a pre-defined or template playing field. The terms "field registration" and "camera calibration" are sometimes used interchangeably, but they have distinct purposes [107]. Sports field registration estimates a homography between the 3D sports field plane and the image, which is limited to applications within the field plane. In contrast, camera calibration provides a mapping between the entire 3D world and the image, making it suitable for comprehensive 3D applications. By achieving precise field registration, analysts can create a reliable spatial context for tracking player movements and game events, enabling a consistent frame of reference throughout the analysis. This chapter will primarily introduce field registration techniques for sports analytics.

Tracking includes both the detection and association of relevant objects within video frames. In sports, this involves first detecting players, the ball, and other key elements using advanced algorithms like neural network approaches that accurately locate and classify these objects, even under varying lighting conditions and occlusions. Once detected, for example, some tracking algorithms predict the positions of these

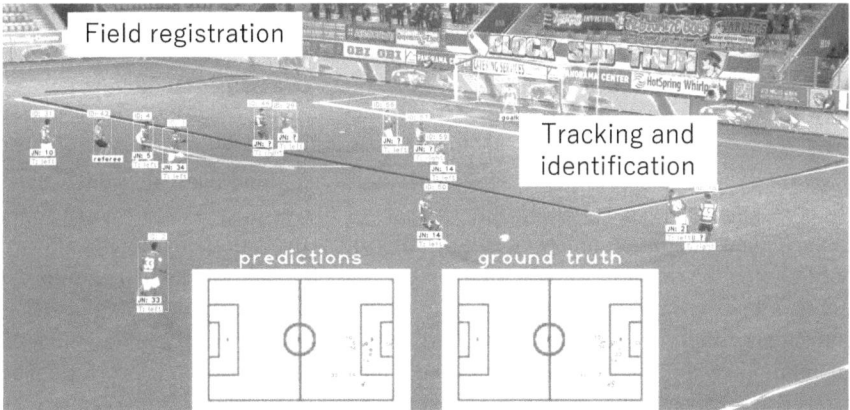

Fig. 2.1 An illustration of field registration, tracking, and identification in a soccer match used in a game state reconstruction competition [150]. The figure highlights how positional data is aligned with a pre-defined field model, and players are tracked and identified through their jersey numbers. The minimaps at the bottom show the predicted and ground truth positions on the minimaps. This figure utilizes the image with permission from the SoccerNet Community

objects in future frames based on their current trajectory and velocity. This process of detection and tracking is important for analyzing player movements, including speed and interactions with other players and the ball.

Re-identification addresses the challenge of recognizing the same player across different camera views or after occlusions. This process ensures continuity in tracking individual players throughout the game. Player re-identification is important for automatic video processing in sports, leveraging distinctive features of the player's image, such as jersey color and number, or feature vectors obtained from various feature extraction methods. By using these features, reliable re-identification can be achieved, which is particularly important for long-duration tracking. Deep learning models are often employed to learn and extract these distinctive features, allowing for accurate re-identification and ensuring reliable tracking of individual players throughout the game.

Action recognition and detection is the process of identifying specific actions or events in sports footage, such as passes, shots, tackles, or goals. This involves analyzing sequences of frames to detect patterns that correspond to particular actions, with rule-based or deep learning models commonly used to recognize these actions based on spatial and temporal features extracted from video data. In addition to recognizing actions in each video, related tasks like action spotting and temporal action segmentation broaden the scope of analyzing sports footage. Action spotting focuses on pinpointing the exact time at which a particular event occurs. It provides precise timestamps for key actions, which is particularly useful for generating automated highlights and applications requiring exact event timing. Temporal action segmentation or localization involves dividing the video into segments based on the occurrence of actions, identifying the start and end times of actions, and offering a more detailed understanding of the duration and sequence of events. By considering action recognition, action spotting, and temporal action segmentation, advanced systems can automate the annotation of key events in a game, facilitating detailed performance and tactical analysis.

Pose estimation and tracking involves determining the precise positions of a player's body joints from video footage. By capturing and analyzing the skeletal structure of players, it allows for the analysis of player postures, movements, and biomechanics. In pose estimation, algorithms such as using deep learning models are used to map the locations of key joints (e.g., shoulders, elbows, knees) within each frame. This data can then be used to assess techniques, identify potential injury risks, and enhance training programs. Pose estimation is especially valuable for both team and individual sports, where body mechanics are important to enhance performance.

By focusing on these essential components, computer vision systems can provide a robust framework for sports analytics. Each element plays a critical role in transforming raw video footage into meaningful data that can be used to derive insights into player performance and team tactics.

2.1.2 Importance and Challenges of Data Acquisition in Sports

Accurate and reliable data acquisition is the cornerstone of effective sports analytics. The quality of the insights and analyses derived from computer vision techniques heavily depends on the quality of the data collected. In sports, data acquisition involves capturing various types of visual and positional information from games and practices, which are then used to inform decisions and strategies.

However, acquiring high-quality data in sports presents several challenges:

- **Complexity of movements**: Sports involve fast-paced and complex movements, making it difficult to capture precise data by hardware (cameras) like motion blur. In addition, players on the same team often wear similar uniforms, sudden changes in movement direction, and moves with intentional physical contact (increases the frequency of occlusion), complicating accurate tracking, identification, and other techniques.
- **Environmental factors**: Weather conditions including lighting, camera angles, and obstructions can significantly affect the quality of the captured footage. Developing methods that are robust to various environmental factors is a major challenge because it is nearly impossible to ensure consistent and clear visuals across different stadiums and sports.
- **Volume of data**: Sports events generate large volumes of data, including video footage, tracking data, event data, pose data, and other sensor information. Managing and processing this data efficiently is critical for timely analysis and feedback.

- **Integration of multiple data sources**: Combining data from various sources, such as cameras, GPS devices, and wearable sensors, to create a cohesive dataset can be complex. Ensuring the synchronization and compatibility of these data streams is essential for accurate analysis.
- **Privacy and ethical considerations**: Collecting detailed data on athletes raises privacy and ethical concerns. Ensuring that data collection practices comply with legal and ethical standards is important for protecting the rights and well-being of all athletes, with particular attention needed for children.

Despite these challenges, advancements in computer vision technologies continue to enhance the capabilities of sports analytics. By addressing these issues and leveraging state-of-the-art techniques, researchers and practitioners can obtain high-quality data that drives better performance and tactics.

In the following sections, specific technologies and datasets used in computer vision for sports analytics are explored, highlighting their applications, methodologies, and impact on the field. In particular, these sections emphasize how open-source datasets benefit the scientific community by driving research in this field.

2.2 Description of Key Computer Vision Elements for Sports Analysis

This section provides a detailed examination of the work carried out in each key task of computer vision used in sports analysis, including the available datasets and developed methods. We discuss how these elements, including field registration, tracking, re-identification, action recognition and detection, and pose estimation, are essential for extracting meaningful insights from sports footage. Each of these components plays a crucial role in the analysis pipeline, contributing to precise data acquisition for sports analytics. Through a detailed examination of these elements, the advancements in computer vision technology and their impact on sports analytics are introduced.

2.2.1 Field Registration

Field registration is an important process in sports analytics, aligning video footage with a pre-defined playing field to ensure that positional data accurately corresponds to real-world coordinates. This alignment allows for precise tracking of player movements, forming the foundation for various analytical tasks. This section first introduces available datasets for field registration, and then different approaches to field registration are explored, including camera calibration techniques, recent keypoint-based and keypoint-less approaches, and strategies for handling multiple camera views and occlusions.

Datasets for Field Registration

To develop and evaluate field registration methods, researchers have shared and utilized various datasets annotated with such as homographies and geometric field elements in sports fields. These datasets are important for benchmarking and advancing the accuracy of field registration techniques. Here, a summary of some significant datasets used in sports analytics is presented, highlighting their scope, annotations, and accessibility. As of 2024, the datasets have been comprehensively summarized by Magera et al. [107]. Note that the following datasets include field markings, homographies, or pinhole models. Field markings include geometric field elements such as lines and circles in sports fields. Homography is a simpler transformation suited for 2D planar surfaces, while the pinhole model provides a complete 3D mapping that accounts for the camera's intrinsic and extrinsic parameters, enabling more complex spatial analyses.

In soccer, the WorldCup 14 dataset includes 395 images annotated with homographies and is openly available. It has been extensively used in studies such as [29, 67], making it a popular choice for evaluating field registration methods. Similarly, the TS-WorldCup dataset provides 3,812 soccer images with homography annotations and is also publicly accessible [33]. The SoccerNet-calibration dataset [35] is another notable resource, offering 21,132 soccer images annotated with field markings. This dataset is publicly available and has been employed in several studies, such as [165]. The CARWC dataset contains 4,207 soccer images with homographies and is openly available [39].

In ice hockey, the SportLogiq dataset reportedly had 1.67 million images, although the specifics of the annotations are not disclosed, and the dataset is not publicly accessible [67, 81]. For multi-sport applications, the SportsFields by Amazon dataset includes 2,967 images annotated with homographies but is also not publicly available [122]. For volleyball, a dataset containing 470 images with homographies was reported [29], but the annotation data is not publicly available.

In basketball, DeepSportRadar and 3DMPB, offer 728 and 10,000 images respectively, both annotated using the pinhole model and are publicly available [72, 170]. The College Basketball dataset consists of 640 basketball images with homographies and is not publicly accessible [145]. Lastly, the Athletics dataset provides 10,000 images with pinhole model annotations and is also publicly accessible [13].

These datasets allow the development for the development of field registration algorithms, providing diverse and annotated images that facilitate the creation and evaluation of reliable camera calibration methods. However, the varying nature of annotations, limited dataset sizes for certain sports, and restricted access to some datasets pose challenges in achieving consistent and comparable results across different studies.

Traditional Field Regisration Approaches

Field registration has been basically considered as a homography estimation problem, whereby corresponding features or keypoints are identified between the image and the field model, and the mapping between them is computed using techniques such as RANSAC (Random Sample Consensus) [48] and DLT (direct linear transformation) [60]. These methods rely on detecting specific points or features on the field, such as corners and line intersections, to compute the homography or camera parameters required for field registration.

Traditional approaches primarily focus on estimating homography, which only provides a mapping between the 2D field plane and the image. However, it is challenging to account for camera distortions, changes in perspective, and variations in camera parameters, leading to less accurate and reliable results. Therefore, camera calibration is also important in field registration, and it is important to recognize that the two tasks complement each other. Some works have explored camera calibration to enable the projection of non-planar points, such as those belonging to goal posts or crossbars (e.g., [27, 35, 150]). This involves estimating the camera's intrinsic parameters (focal length, principal point) and extrinsic parameters (rotation, translation) using the known dimensions of the sports field as a calibration rig.

Deep Learning Approaches in Field Registration

Despite the precision offered by traditional field registration and calibration methods, they often rely on manual feature selection and parameter tuning, making them less adaptable to varying conditions. Deep learning approaches have emerged as a powerful alternative, leveraging large datasets and advanced algorithms to automate and enhance the field registration process, offering improved robustness and efficiency. Keypoint-based approaches using deep learning have focused on either directly predicting an initial homography matrix [81, 163] or seeking the optimal matching homography within a reference database of synthetic images with known homographies or camera parameters [145, 147, 188, 189]. Other methods exploit the temporal consistency between subsequent video frames [38, 122] to refine the homography estimates. Also, geometry-based 3D sports field registration has been proposed [58], which is currently the best open-source method for soccer field registrations [58]. By employing classical camera calibration techniques such as the DLT algorithm and RANSAC, the method achieved superior performance in both multiple- and single-view 3D camera calibration.

Keypoint-less approaches do not rely on detecting specific points or features on the field. Instead, they use more generalized features, such as lines or regions, to achieve field registration. TVCalib [165] uses a differentiable objective function to learn camera pose and focal length from segment correspondences. By using segment localization and an iterative calibration module, this approach minimizes reprojection errors and performs well even with broadcast soccer videos [37, 58].

Common challenges in field registration

A common challenge in field registration is the presence of partial occlusion of the court in broadcast videos. To address this, recent methods have incorporated deep learning-based semantic segmentation [67, 122, 188] or edge detection [147] to extract relevant features from the field. For another challenge, current approaches still largely overlook the complexities introduced by camera lens distortions and non-linearities in the mapping between video footage and real-world coordinates. Some approaches have focused on leveraging the geometric properties of sports fields to generate keypoint grids, enabling robust camera calibration with minimal refinement using DLT and RANSAC algorithms [58]. Investigating new models and algorithms that can accurately account for these factors will be important in further enhancing the precision and reliability of field registration. Such advancements in field registration contribute significantly to the overall effectiveness of computer vision applications in sports analytics, facilitating detailed and accurate analysis of player movements.

2.2.2 Tracking

Researchers have proposed various methods for tracking players and the ball in team sports. Tracking involves detecting and following the movements of players and the ball across successive video frames (the latter is called multi-object tracking (MOT)), enabling to generate accurate positional data, which serves as the foundation for further analytical tasks. Various traditional and deep learning approaches have been developed to enhance the accuracy and efficiency of tracking in sports. Previous surveys [110, 133] summarized various approaches that combine background subtraction, multi-camera triangulation, and Kalman filters to track player movements on the pitch, and here, a more brief but comprehensive review including recent approaches is presented.

Datasets for Tracking

In the sports domain, pioneering work in sports tracking datasets has provided soccer player location data using multi-view video cameras, offering valuable resources for player tracking and analysis [44]. Additionally, datasets featuring 2K panorama monocular and multi-view videos combined with LPS (local positioning system) data have further enhanced the ability to track player movements accurately [131]. For broader applications, large broadcast video datasets such as SoccerNet [36, 150] and SoccerDB [82] have been made publicly available. In other team sports, datasets like APIDIS and SPIROUDOME for basketball [41, 104], handball [19], as well as volleyball video datasets [75], provide comprehensive video data for tracking players.

Recent datasets such as SoccerNet-Tracking [36] and SportsMOT [40] (soccer, basketball, and volleyball) utilize unedited main camera and broadcast footage. These datasets often require additional processing for image registration and handling zoom. SoccerTrack [142] offers full-pitch tracking using drone and fisheye cameras and has been recently extended to the TeamTrack dataset, which provides multi-sport (including basketball and handball) MOT in full-pitch videos [141] as illustrated in Fig. 2.2. Furthermore, virtual environments such as Google research football (GFootball) [89] enable the generation of synthetic camera and location data, providing a controlled setting for developing and testing tracking algorithms. These advancements in tracking datasets significantly democratize accurate tracking for sports analytics, as well as allow benchmarking algorithms and comparing their performance in a standardized environment.

Object Detection

For accurate tracking of players and balls, object detection is the initial step, involving the identification of relevant objects within a video frame. Traditional approaches utilize techniques like background subtraction, which helps distinguish players and the ball from static parts of the field. This method isolates moving objects by subtracting the background image from the current frame. Feature-based methods such as

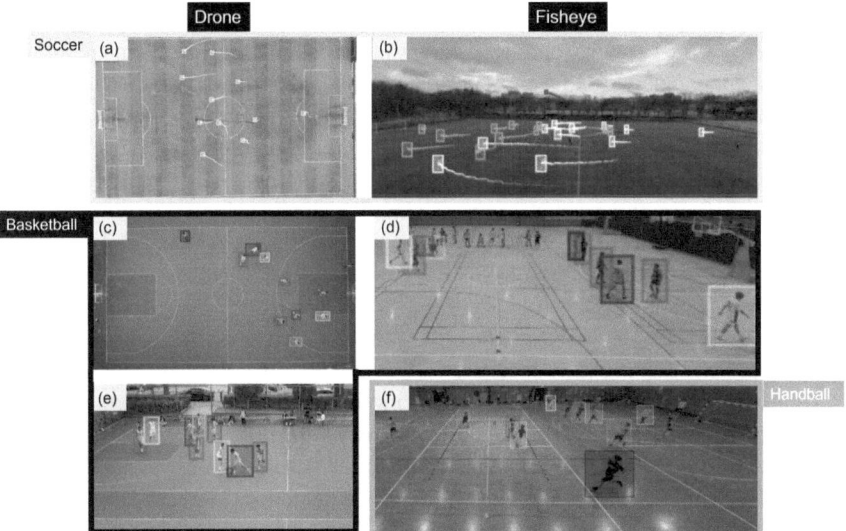

Fig. 2.2 An illustration of TeamTrack dataset [141]. It offers full-pitch tracking using a drone (**a**, **c**, **e**) and fisheye (**b**, **d**, **f**) cameras in multi-sport MOT in full-pitch videos. The dataset covers soccer (**a**, **b**), basketball (**c**, **d**, **e**), and handball (**f**) with different camera perspectives, enhancing the diversity and applicability of tracking methodologies. From the arXiv version of [141], the figure is licensed under CC-BY 4.0

Histogram of Oriented Gradients (HOG) use gradient information to detect players and the ball [32, 106]. Edge detection methods, such as the Canny edge detector and Sobel filtering, identify the edges and contours of players and the ball (e.g., [4, 43]. Recently, deep learning approaches have been used such as in RetinaNet [99], CenterNet [45], and YOLO (You Only Look Once) [136]. YOLO, in particular, has seen numerous iterations and improvements, with newer versions continually being developed to enhance detection accuracy and speed. However, off-the-shelf algorithms like YOLO tend to over-detect non-athlete individuals, such as spectators or staff around the field. Therefore, fine-tuning these models on sports-specific datasets is important to ensure that detection focuses primarily on the athletes on the field.

Tracking-by-Detection Approach

Once objects are detected, the next step is to associate the detections across successive frames, which is called the tracking-by-detection approach. Traditional tracking methods include point tracking techniques of players and the ball such as the Kalman filter (e.g., [61]). Contour tracking, silhouette tracking, and graph-based tracking involve tracking the contours or boundaries of players and the ball using active contour models (e.g., [91]), employing shape analysis techniques to match and track silhouettes (e.g., [117]), and representing player positions as nodes on a graph with trajectories as edges (e.g., [47]), respectively.

In recent MOT, associating detected objects across frames to maintain consistent identities is a key challenge, particularly in scenarios involving occlusion or rapid motion. Two primary approaches for addressing this association problem are motion-based and appearance-based techniques, each with its own strengths and limitations.

Motion-based methods rely on predicting the future positions of detected objects using their past trajectories. SORT (Simple Online and Realtime Tracking) [18] have gained popularity for their speed and simplicity, which uses a combination of the Kalman filter for motion prediction and the Hungarian algorithm for data association. Building on this, many algorithms have been developed such as ByteTrack [191] and OC-SORT (Observation-Centric) [24], BIoU (the buffer of two overlapping boxes) [183], EIoU (expanding the IoU according to different scales of expansion) [74]. BIoU and EIoU also demonstrate the effectiveness of MOT on the SportsMOT [40] and SoccerNet-Tracking [36] datasets.

Appearance-based methods complement motion-based techniques by using visual features to distinguish and associate objects. In tracking, this approach is particularly important for maintaining consistent identification across frames, often leveraging re-identification features to associate tracklets or detections accurately, which also appears in the following subsections. DeepSORT is a pioneering method that incorporates deep visual features for object association [174]. Such deep learning approaches have been used for association via feature extraction, such as by convolutional neural networks (CNNs) [62, 194] and transformers [63, 148]. However, these methods face challenges when dealing with objects that have similar visual characteristics or are frequently occluded (e.g., [154]). Despite these challenges, appearance-based approaches provide valuable complementary data, enhancing the overall robustness of tracking systems when combined with motion-based techniques.

Unique Challenges for MOT in Sports

MOT in sports environments presents significantly greater challenges. This complexity arises due to the unique characteristics of sports, such as the rapid and unpredictable movements of athletes, the visual similarity among players within the same team, and the increased occurrence of occlusions due to the dynamic nature of the sport. Instead of a tracking-by-detection paradigm, in domains other than sports, end-to-end tracking methodologies integrate object detection and tracking into a single process, potentially improving performance by handling both tasks concurrently. For instance, Tracktor [17] leverages frame redundancy to streamline data association, while Neural Solver [20] and DeepMOT [180] utilize neural and Siamese networks to improve tracking accuracy. Transformer-based models like DETR [26] have also been adapted for tracking, as seen in Trackformer [114] and TransTrack [155].

Despite their potential, end-to-end tracking approaches are not widely used in current sports MOT. This may be due to such as less amount of annotation data and the complex nature of sports environments including severe occlusions and others mentioned above. Several researchers have made notable contributions to address such challenges in various sports. For example, in football, tracking accuracy was improved through simultaneous field and player localization [108]. MV-Soccer [109] leveraged motion vectors to enhance real-time detection, instance segmentation, and tracking of soccer players. In basketball, human pose information and actions were utilized as embedding features for player tracking [7]. In highly occluded scenarios such as wide-view basketball from the TeamTrack dataset [141], Hu et al. [71] developed Basketball-SORT to reacquire long-lost IDs based on specific basketball scene characteristics. In ice hockey, Vats et al. [172] combined team classification and player identification techniques to enhance tracking. In multiple sports, including basketball, volleyball, and soccer, Huang et al. [73] integrated OC-SORT with appearance-based post-processing. In summary, these advancements are important for sports analytics, enabling detailed analysis of player movements.

2.2.3 Re-identification

Player re-identification (Re-ID) is an essential process in sports video analysis, aiming to identify and distinguish each player across different video frames and camera angles. This process is critical for automated video processing tasks such as tracking player movements, generating highlights, and analyzing player behavior. Player Re-ID leverages various features from players' images, including jersey color and number, and more sophisticated feature vectors obtained from deep learning models.

Datasets for Re-ID in Sports

Datasets for Re-ID in non-team sports often originated from simple identification tasks, where the primary goal was to distinguish athletes using features like bib numbers or facial recognition. The RBNR Dataset [15] is specifically tailored for marathon events, containing 217 color images with annotated ground truth bib numbers, which are important for training and testing Re-ID models in long-distance running scenarios. Similarly, the Marathon RBN Dataset [129] includes 9,706 images captured in daylight, providing ample data for re-identifying marathon runners. Another significant dataset in this category is the TGC20ReId Dataset [130], which addresses the Re-ID problem in an ultra-running sport event. This dataset comprises 4,373 images of runners captured under varying lighting conditions (day and night), across 2–5 different locations, offering a novel approach to player Re-ID in more challenging outdoor environments. Recently, a re-ID model and evaluation dataset for runners [159], focusing on consistently tracking athletes across frames without relying solely on known identity markers, is provided, which is described later.

In contrast, team sports datasets for Re-ID often involve recognizing players by their jersey numbers or other visual attributes. The Synergy Re-Identification Dataset [170] provided by DeepSportradar, includes image crops of basketball players, referees, and coaches from short game sequences, with a training set of 8,569 images, and various query and gallery images for testing and challenges. The Hockey Dataset [86], which includes legible images of players' jersey numbers from university ice hockey games, and the McGill Hockey Player Tracking Dataset [192], spans multiple games and provides extensive data for player Re-ID in hockey. Furthermore, the SoccerNet Re-Identification Dataset [35] provided 340,993 player thumbnails extracted from broadcast videos of 400 soccer games across six major leagues. This dataset is designed to re-identify soccer players across multiple camera viewpoints during games, offering a comprehensive resource for evaluating and training Re-ID models in team sports scenarios. Similarly, the SoccerNet Game State Reconstruction dataset [150] offered a video-based re-identification (through time) and identification (jersey number) setup, which is described later. These datasets present a range of challenges including similar uniforms, dynamic movements, and varying lighting conditions, essential for advancing player performance analysis in sports analytics.

Image-Level Re-ID

Also in sports Re-ID, image-level methodologies have been first developed. For example, early research explored visual local features related to the faces of soccer players [11]. Jersey numbers are distinctive and easily recognizable, making them reliable identifiers. For instance, jersey colors and numbers have been utilized for player detection in basketball and soccer, demonstrating the utility of these features in sports analysis [79, 120].

However, relying solely on jersey numbers is insufficient due to various challenges such as occlusions, varying camera angles, and low resolution. To address these challenges, advanced techniques involving feature extraction from deep learning models have been developed. For example, the use of deep convolutional representation and multi-scale pooling for part-based player identification has shown promise [143]. Additionally, body feature alignment techniques have been employed in soccer to improve Re-ID accuracy [2].

Advanced deep learning methods have further enhanced player identification in various sports. For instance, techniques like constrastive learning and the use of CLIP (Contrastive Language-Image Pre-Training) models have been employed for more robust player identification in hockey and basketball [59, 87, 171]. Multi-task learning frameworks have also been developed for joint Re-ID, team affiliation, and role classification in soccer visual tracking [111]. The Attention-Aware Multiple Granularities Network (A2MGN) captures discriminative features from different granularities using multiple branches, including a global branch, part-based branches, and an attention-aware branch [5]. Additionally, an enhanced Swin Transformer has been developed for soccer player Re-ID, leveraging advanced deep learning architectures to further boost accuracy [3].

More Realistic Setting

Here, a more realistic Re-ID setting can be considered. In sports, Re-ID involves not only recognizing players based on individual images but also tracking them over sequences of frames, known as tracklets. Tracklet-level Re-ID [86] is important for long-term tracking and analysis, addressing challenges such as motion blur, occlusions, and varying camera angles. The approach involves aggregating information across multiple frames to maintain consistent player identification. To solve these issues, a robust pipeline was proposed that starts with the main subject filtering to isolate frames where the player is not occluded. A legibility classifier identifies frames with clear jersey numbers, and a scene text recognition system is then used to recognize the jersey numbers in these frames [86]. The final step aggregates these frame-level predictions into a tracklet-level prediction, enhancing the accuracy and reliability of jersey number recognition in sports videos.

Also in [150], athlete Re-ID is evaluated at the video level rather than just the image level. The Re-ID process involves tracking athletes throughout the game by integrating multiple attributes such as their role, team, and jersey number. This comprehensive approach addresses challenges like occlusions, varying camera angles, and partial visibility of jersey numbers by leveraging feature extraction from deep learning models. This game state reconstruction [150] is described in Sect. 2.3.

Another topic is the open-world problem setting. Most studies on player Re-ID focus on the closed-world setting, which relies on improving the performance of feature extractors using pre-prepared image datasets with sufficient labeled data [184] (Fig. 2.3 left). However, this approach is limited for general sports video processing due to the high cost of labeling image datasets from videos and the need to

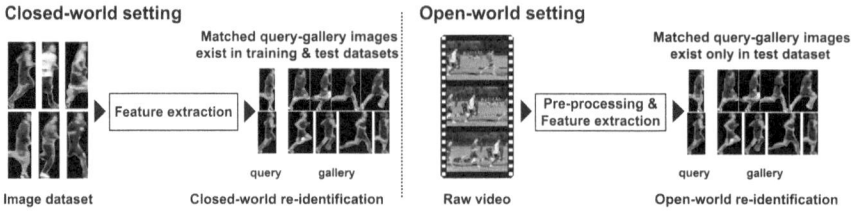

Fig. 2.3 The open-world setting for runner Re-ID. The left illustrates the closed-world setting, where pre-labeled images can be used for training. The right depicts the open-world setting, where raw video data is processed directly, allowing for the identification of players not present in the training dataset. This approach is particularly suited for general sports video processing and daily practice scenarios. The pictures are the original

identify players not included in the dataset. Consequently, the open-world setting, which involves directly processing raw images or videos, including person image extraction and feature extraction, is more suitable for real-world applications [184] (Fig. 2.3 right). This setting does not require manual data processing and is particularly beneficial for daily practice videos where cost constraints are significant with video measurement and processing costs. In this open-world setting, Suzuki et al. [159] addressed this by tracking runners, detecting their shoes, and extracting image features through an unsupervised method with a mix of global and local features. In summary, such advancements in Re-ID technology contribute significantly to the automation and accuracy of player performance analysis in sports analytics.

2.2.4 Action Recognition and Detection

Action recognition and detection in sports videos is a complex field involving various tasks such as action recognition, action spotting, temporal action segmentation, and other specialized tasks. These tasks are crucial for analyzing sports events, understanding player behaviors, and automating video analysis processes. Action recognition usually identifies one action class describing the whole video clip, but action detection or temporal action segmentation/localization refers to the task of identifying the times of actions within a video. In action detection, action spotting focuses on pinpointing the exact time at which a particular event occurs. Temporal action segmentation/localization identifies the start and end times of actions based on the occurrence of actions. By considering action recognition, action spotting, and temporal action segmentation, advanced systems can automate the annotation of key events in a game, facilitating detailed performance and strategy analysis. Here, datasets for action recognition and detection are first introduced, then the details of these tasks and the proposed methods are explained.

Datasets for Action Recognition and Detection

Researchers have established numerous benchmarks in the sports domain to address challenges to explore promising application prospects. For video action recognition, datasets such as UCF Sports Action [138] and Diving48 [96] provide extensive collections of action clips for training and evaluation. Group action recognition, which involves tagging each video clip with a group action label, utilizes datasets like basketball [182], volleyball [76], ice hockey [9, 10], and collective sports [187] highlighting the interactions and coordinated movements among team members.

Action spotting tasks were introduced by the SoccerNet dataset [37, 42, 53], which focuses on identifying specific moments within sports footage. Temporal action segmentation, which aims to detect and segment continuous actions over time, is addressed using datasets like diving [179], gymnasitics [146], and figure skating [50, 102]. MultiSports dataset [97] including aerobic gymnastics, basketball, soccer, and volleyball is also commonly used for this purpose. Additionally, SportsHHI dataset [176] provides interaction definitions and annotations to explicitly explore human-human interaction. These datasets collectively advance the field of sports analytics by providing rich sources of data for various analytical tasks.

Action Recognition Studies

Action recognition involves identifying and classifying a specific action within a video segment [66], assuming that the temporal boundaries of the actions are already known (or rather not of interest) and focusing solely on recognizing the type of action being performed. In the context of team sports, it is essential to focus on group action recognition, which is critical for understanding team dynamics and strategies. A comprehensive review [175] categorizes methods for group activity recognition (GAR) into traditional approaches based on handcrafted features and more recent methods utilizing deep learning. Traditional approaches are divided into top-down methods, which analyze global group motion and interactions, and bottom-up methods, which focus on individual actions and their aggregation. Deep learning methods are discussed in terms of hierarchical temporal modeling, relationship modeling, attention modeling, and unified frameworks, highlighting both the progress and challenges in the field. Many studies use the volleyball dataset [76], and recent research has explored Transformer architectures using pose information and person attributes (e.g., [94, 119]).

Specific event recognition in sports has seen advancements in various areas, addressing the need for automated systems to assist in decision-making processes. In soccer, automated offsides detection has been developed to enhance the accuracy and speed of rulings during matches [168]. Additionally, a video assistant referee (VAR) system has been proposed for making automated decisions from multiple camera views, improving the overall fairness and consistency in soccer games [64]. In rugby, the detection of dangerous tackles has been explored to enhance player safety and game integrity [124]. Similarly, in athletics, methods for the automatic detection of faults in race walk have been developed, providing reliable and objective assessments of athlete performance [156, 157]. Furthermore, edge error detection in figure skating has been introduced to support judges in identifying technical mistakes, thereby contributing to more accurate scoring [162]. These innovations emphasize the growing role of automated event recognition systems in improving the accuracy, safety, and fairness of sports competitions.

Action Detection and its challenges

In sports analytics, it is crucial to identify key actions within a video. Action spotting, focusing on identifying specific moments or key events within a video, is essential in terms of sports analytics. Unlike action recognition, which deals with pre-segmented actions, action spotting involves locating the precise time points where actions occur. This task is particularly relevant in sports, where fast-paced actions need to be pinpointed accurately. Action spotting and the following temporal action segmentation share common techniques, but the choice between them depends on the characteristics of the sport. For instance, action spotting is often preferred for identifying discrete events, such as passes in soccer or jumps in figure skating, where the action is best captured at specific moments.

Recent action spotting techniques benefit significantly from end-to-end (E2E) approaches, compared to feature-based approaches (e.g., [34, 55, 195] due to their ability to simultaneously learn features and make predictions in a unified framework. For instance, E2E-Spot [69] combines the steps of finding important features and making predictions into one unified model, making the system more efficient and accurate. Building upon these concepts, T-DEED (Temporal-Discriminability Enhancer Encoder-Decoder) [177] further enhanced the approach by refining token discriminability and handling multiple temporal scales. In particular, T-DEED [177] has achieved both novelty and high performance, even securing first place in the SoccerNet Ball Action Spotting 2024 competition. In addition, a recent open-source library aims to gather all methods into a single framework [16]. While action spotting can be done manually by a team member or outsourced if budget allows, it is a labor-intensive process that not every team can manage. Nevertheless, the insights gained from recognizing these events are essential for game analysis, providing fundamental information. Therefore, improving the accuracy of action spotting and the following temporal action segmentation is an essential step toward automating data acquisition in sports analytics.

Temporal action segmentation (TAS) involves dividing a video into segments, each corresponding to a distinct action or event. This task is important for creating structured representations of untrimmed videos, enabling detailed analysis and understanding of continuous video streams. TAS methods often use both proposal generation and classification stages to identify and label each segment accurately. Although there are several datasets available for TAS in sports as described above, TAS for sports has not been extensively studied (often focusing on cooking procedures in kitchens such as [88] and assembly procedures for furniture and toys such as

[12]). This might be due to the inherent complexity and variability of sports actions, which are often more dynamic and context-dependent than those in kitchen or assembly tasks. TAS models previously had frame-based approaches (e.g., [46, 186]) that rely on temporal convolution networks or tranformers to refine feature representations over multiple stages. On the other hand, two-stage methods (e.g., [14]) aim to capture long-range temporal information by learning action features in two distinct phases. In sports, the boundaries of events are often not well-defined but TAS has intensively investigated in figure skating [98, 160, 161]. In particular, Tanaka et al. [161] used the FACT framework, which integrates frame and action features through bidirectional cross-attention [105], to perform TAS in figure skating jumps by leveraging 3D pose data. Despite these challenges, there is significant potential for TAS in sports (reviewed about soccer [144]) because it requires more in-depth knowledge of the specific sport, and automatic acquisition of such fundamental information is highly anticipated and essential from a sports analytics perspective.

2.2.5 Pose Estimation

Pose estimation in sports presents unique challenges due to the dynamic and complex movements involved, the limitations of camera setups, and the frequent occlusions and contact between players. Sports activities often require capturing fast, intricate motions, which is a significant challenge compared to existing datasets. Additionally, sports scenes sometimes involve moving, low-resolution, and complicating pose estimation efforts. Team sports also introduce substantial occlusions as players frequently come into contact, and the similar appearance of players wearing matching uniforms makes it difficult to distinguish between individuals using appearance-based methods. This introduces the necessity of specialized datasets, advancements in both 2D and 3D pose estimation techniques, and solutions to these challenges.

Datasets for Pose Estimation in Sports

2D pose estimation in sports often relies on well-annotated datasets to train and evaluate models. Some notable 2D pose datasets include the LSP dataset, which contains 1,000 images specifically curated for human pose estimation in sports settings [84]. The Sport Image dataset provides 1,300 images focused on various sports activities, facilitating detailed analysis of athletic poses [173]. The MPII Human Pose dataset is more extensive, offering 25,000 images capturing a wide range of human activities, including sports [6]. OCHuman dataset, with 4,700 images, emphasizes occluded human poses, which are common in dynamic sports scenes [190]. The COCO-WholeBody dataset is the most comprehensive, with 250,000 images covering the whole body, including detailed annotations for face, hands, and feet [83]. Additionally, the 3DSP dataset offers 4,000 images specifically annotated for soccer player poses [185] and KTH Multiview Football Dataset II [85] provides 800 frames pose data captured from 3 views. These are built upon extensive manual effort. The DeepSportLab dataset[1] [52] can be utilized to train a unified model that simultaneously predicts ball detection, player instance segmentation, and pose estimation in team sports scenes, demonstrating its effectiveness specifically in basketball contexts (672 images of professional basketball games captured from 29 arenas).

3D pose estimation requires datasets that provide depth and spatial coordinates to capture the complexities of athletic movements. Creating a 3D pose dataset requires the use of multiple cameras, or for more precise data, an optical motion capture system. The ASPset-510 dataset offers 330,000 frames focused on various sports, providing detailed 3D annotations to support sports-related pose estimation [121]. The Sportspose dataset includes 1.5 million frames specifically designed for sports pose estimation, facilitating advanced research in this area [77]. The AIST++ dataset is one of the largest, containing 10.1 million frames of dancing sequences, which helps in understanding complex motion patterns [93]. The Runner dataset, with about 20,000 frames, is tailored for running activities [158]. Recently, figure skating jump dataset [161] was published including both 3D pose data and video data from 12 viewpoints and 78,000 frames. However, in general, there is a shortage of 3D pose datasets specialized in specific sports movements, excluding the above, due to the fact that it is hard to capture 3D pose ground truth data on the field (compared to a controlled lab environment). These datasets are essential for developing and refining pose estimation models in sports, enabling accurate analysis and understanding of athletic performance.

2D Pose Estimation

Top-down and bottom-up approaches are the two main strategies for 2D human pose estimation. The top-down approach in 2D human pose estimation has evolved significantly over the years, demonstrating impressive advancements in accuracy and efficiency. This approach typically involves a two-stage process: first detecting individual persons in an image, then estimating the pose for each detected person. DeepPose [167] pioneered the use of deep neural networks for pose estimation, formulating it as a DNN-based regression problem. The High-Resolution Network (HRNet) [153] further improved upon this by maintaining high-resolution representations throughout the entire process, leading to more accurate and spatially precise keypoint predictions. Most recently, ViTPose [181] demonstrated that even a simple Vision Transformer-based model, without complex modules or CNN fusion, can achieve competitive results in human pose estimation. These top-down approaches are often more accurate than the following bottom-up approach but computationally expensive and dependent on detection quality.

Bottom-up approaches for 2D human pose estimation have also evolved significantly over the years, offering efficient solutions for multi-person pose estimation.

---

[1] https://ispgroup.gitlab.io/code/deepsportlab/.

DeepCut [132] pioneered the use of deep learning for jointly detecting and associating body parts. OpenPose [25] further advanced the field by introducing part affinity fields for real-time multi-person pose estimation. HigherHRNet [31] improved upon previous methods by introducing a scale-aware representation learning approach, particularly effective for detecting poses of people at different scales. Bottom-up approaches are often more efficient and scalable than top-down approaches but can struggle with accurate keypoint grouping and may sometimes yield lower overall accuracy. More recently, end-to-end approaches that blur the line between bottom-up and top-down methods have emerged (e.g., [100, 149]). These end-to-end methods try to address the challenges of multi-person pose estimation, such as occlusions and varying scales, by leveraging the global context provided by transformer architectures.

3D Pose Estimation

3D pose estimation has seen significant advancements through various approaches, including direct estimation and 2D-to-3D lifting. Here, monocular 3D pose estimation methods are first introduced, and then multi-view ones are described. Direct estimation methods aim to predict 3D poses directly from 2D images without intermediate 2D pose estimation (A seminal work is [164]). This approach, while innovative, often struggles with depth ambiguities and occlusions. In contrast, 2D-to-3D lifting methods first estimate 2D poses and then lift these 2D keypoints to 3D space (pioneering work is [112]). Recently, transformer-based models have been used in sports 3D pose estimation. For example, StridedTransformer-Pose3D [95] was used in figure skating [162], and MotionAGFormer [113] was used for soccer broadcast videos [185] in our group as shown in Fig. 2.4.

Multi-view 3D pose estimation has gained significant attention due to its ability to handle complex scenarios and occlusions. To address these issues, for example, a learnable triangulation of human pose method [78] and a method simultaneously reasoning about multiple individuals' 3D body joint reconstructions and associations in space and time [135] were proposed, which allow for end-to-end training and improved accuracy. In sports, a method for learning monocular 3D human pose estimation from multi-view images is proposed, leveraging view consistency and a small set of labeled data to enable pose estimation for motions with limited annotations [137]. A fast greedy algorithm for multi-person 3D pose estimation and tracking in sports was also presented, which associates 2D poses across views and generates 3D skeletons to handle challenging sports scenarios [21]. Additionally, a motion-aware and data-independent model for multi-view 3D pose refinement in volleyball spike analysis is introduced, utilizing multi-view relationships and sport-specific motion patterns to improve pose estimation accuracy [103].

Fig. 2.4 A 3D pose estimation example from soccer broadcast videos in our group [185]. The top photo is the broadcast video frame, and the bottom-left and bottom-right images correspond to the 2D and 3D pose estimations, respectively. This figure utilizes the image with permission from the SoccerNet Community

Challenges in Sports Pose Estimation

Pose estimation in sports presents several significant challenges that impact its accuracy and applicability. One major issue is the stringent accuracy requirements of sports biomechanists, who need highly precise joint center positions for their analyses. Current pose estimation models often fail to consistently provide the necessary level of precision, resulting in a mismatch between the goals of computer scientists developing these models and the specific needs of sports biomechanists. Furthermore, the dynamic and complex nature of sports movements, combined with the frequent occlusions and similar appearances of players in team sports, aggravates the difficulty of achieving accurate pose estimations. These challenges necessitate advancements in pose estimation techniques to better serve the needs of the sports science community.

To address some of these challenges, recent research has explored innovative approaches such as unsupervised fine-tuning. Suzuki et al. proposed a system for unsupervised fine-tuning of monocular 3D pose estimation models [158] as illustrated in Fig. 2.5. Their method leverages multi-view estimation to obtain initial 3D joint

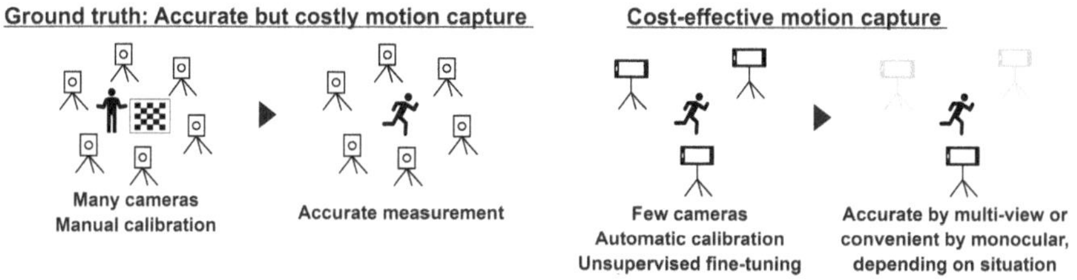

Fig. 2.5 Comparison of motion capture techniques. The figure contrasts ground truth motion capture, which is accurate but costly, with cost-effective motion capture methods [158]. Ground truth motion capture involves many cameras and manual calibration, resulting in accurate measurements. On the other hand, cost-effective motion capture [158] utilizes fewer cameras, automatic calibration, and unsupervised fine-tuning. This approach can be accurate using multi-view setups or convenient using monocular setups, depending on the situation

position estimates, which are then used as pseudo-labels for fine-tuning a monocular model based on [90]. This approach aims to improve the cost-effectiveness and accuracy of sports motion capture by reducing the need for extensive manual calibration and multiple camera setups. In another work, the use of unlabelled data to enhance the robustness and generalizability of pose estimation models was explored [139]. These advancements contribute to the broader field of sports analytics by addressing the specific needs in accurately capturing and analyzing athletic performance.

2.3 Advanced Applications

This section introduces advanced applications of computer vision for sports analytics, highlighting the integration of various techniques to enhance data acquisition and analysis. First, the SoccerNet Game State Reconstruction (GSR) [150] framework is introduced, which is important for sports analytics because field registration, tracking, and Re-ID are combined. Additionally, the advancements in ball state recognition and the prediction and evaluation of actions directly from video are introduced. Finally, the integration of language models with computer vision is examined, demonstrating its potential to bring more context-aware insights to sports analytics.

2.3.1 Integration of Field Registration, Tracking, and Re-ID

The SoccerNet GSR framework integrates field registration, tracking, and identification to achieve a comprehensive understanding of a game's state [150] as illustrated in Fig. 2.1. The process begins with precise pitch localization and camera calibration, enabling the mapping of 2D image coordinates to real-world coordinates on the pitch. This integration ensures that athlete positions are accurately projected onto the field, providing a foundational layer for subsequent tracking and identification tasks. Player detection is carried out using advanced MOT algorithms, which are essential for continuously monitoring player movements throughout the game. The Re-ID process then leverages attributes such as jersey number, team affiliation, and player roles to uniquely identify each athlete, even in scenarios involving occlusions or players with similar appearances.

The SoccerNet-GSR dataset plays a pivotal role in advancing the integration of field registration, tracking, and Re-ID by providing dense annotations that facilitate accurate pitch localization, player positions, and identities. This dataset is instrumental for training and evaluating models, ensuring that the reconstructed game state is precise and reliable. To measure the effectiveness of these models, the GS-HOTA (Game State Higher Order Tracking Accuracy) metric is introduced, which offers a comprehensive evaluation by considering projection accuracy, tracking accuracy, and the quality of player identification. This holistic approach to performance assessment helps in fine-tuning the models to handle the dynamic and complex nature of sports scenarios. Consequently, the integration of these technologies within the SoccerNet-GSR framework significantly enhances the ability to capture and analyze critical data for sports analytics. This comprehensive data acquisition enables deeper insights into player performance, team strategies, and game dynamics, ultimately contributing to improved coaching, training, and broadcasting in sports.

2.3.2 Ball State Recognition

Ball state recognition has seen significant advancements in recent years, particularly in the field of 3D trajectory reconstruction from monocular video. Pioneering work [115] reconstructed 3D soccer ball trajectories from single static camera systems. Building on this, a method for reconstructing 3D trajectories of ballistic basketball shots from monocular videos [28] was developed. More recently, MonoTrack [101], a new approach for reconstructing shuttle trajectories in badminton from the monocular video was proposed, addressing the unique challenges posed by the high-speed and small size of shuttlecocks. Complementing these trajectory reconstruction methods, a technique for 3D localization of balls from a single calibrated image [169] was proposed, further expanding the capabilities of ball state recognition systems. These advancements collectively demonstrate the evolving landscape of ball tracking and trajectory analysis across various sports, utilizing single-camera setups to extract valuable 3D information.

Ball spin estimation in sports is a complex task that can be approached both indirectly and directly. Indirect methods leverage related elements such as human pose and racket movements to infer spin. For instance, in table tennis, human pose information has been utilized to estimate the spin of a ball [140], while robust racket detection methods have been employed to predict ball spin [51]. Direct methods, on the other hand, focus on the ball itself, analyzing its trajectory (e.g., [30]) or observing specific features of the ball. Techniques that track logos or patterns on the ball's surface, such as logo-based tracking (e.g., [56]) and pattern-based approaches (e.g., [49]), provide direct spin measurements. Recent advancements include the use of event cameras in volleyball and tennis [57, 118], which are sensors inspired by the visual system of animals outputting the brightness changes in a scene.

2.3.3 Action Prediction and Evaluation from Videos

Although it is common to perform predictions and evaluations after extracting data, action prediction directly from a video has been extensively explored in contexts such as pedestrian prediction [134] and cooking [1], and it has also found applications in sports for predicting passes, fouls, and the position and orientation of the ball. In soccer, pass prediction has been enhanced by imitation learning techniques, which utilize player coordinates and body orientation data to calculate pass feasibility among teammates, achieving notable accuracy improvements [8]. Additionally, combining trajectory data with video input has significantly enhanced pass receiver prediction accuracy [68]. Foul prediction in soccer has been advanced through the use of estimated poses from broadcast videos, leveraging methods similar to those used for pedestrian intention prediction [80]. Furthermore, volleyball trajectory prediction has been improved using skeletal motion data from the setter player [151], and real-time forecasting of human body motion has been applied to reduce delays in interactive systems [70]. These advancements demonstrate the evolving capabilities and applications of action prediction in various sports contexts.

For action evaluation in sport from video, a method to automatically score Olympic events from video footage was developed, focusing on diving and figure skating performances [128]. This work was extended to assess action quality across multiple sports, including gymnastics vault and ski jumping [127]. Another approach proposed a joint relation graph to assess action quality by modeling detailed joint interactions in sports videos [126]. A system specifically for scoring figure skating performances from video data was also created [178]. Additionally, a combination of rule-based and computer vision approaches for comprehensive and explainable action quality assessment in diving was introduced [125]. In baseball, PitcherNet was developed for analyzing mechanics and performance from video analytics [22]. While earlier works focused on end-to-end neural models, more recent studies have explored hybrid approaches that incorporate domain knowledge and aim for greater explainability. These advancements in automated action quality assessment have significant potential to enhance coaching, judging, and player development across multiple sports disciplines.

2.3.4 Integration with Language Models

Recent advancements in integrating language models with computer vision techniques have opened new avenues for sports analytics, allowing for more nuanced and context-aware insights. For instance, the GOAL dataset [152] leverages videos of football to develop models capable of understanding and interpreting game events. Similarly, the Sports-QA dataset [92] uses various sports videos to create a question-answering system that can respond to queries about the games.

The SoccerNet-caption dataset [116] focuses on generating captions for football games, providing detailed descriptions of game events from video footage. Building on this, SoccerNet-XFoul [65] includes 22,000 questions and answers specifically related to football fouls, enabling the development of models that can answer detailed questions about rule violations in the sport.

In addition to football, other sports have also benefited from these integrations. For example, research on rugby scene classification [123] uses vision-language models to improve the classification of scenes. Another notable work, DanceMVP [193], applies self-supervised learning with transformer text prompting to assess dance performance, demonstrating the versatility and wide-ranging applications of integrating language models in sports analytics.

2.4 Future Directions in Computer Vision for Sports Analytics

The future of sports analytics will be shaped by the integration of multiple modalities and the continuous advancement of computer vision technologies. By combining visual data with other sensory inputs and developing sophisticated machine learning models, the field aims to enhance the depth and accuracy of analysis, providing comprehensive insights into player performance and team tactics.

2.4.1 Integration of Multiple Modalities

The integration of multiple modalities, such as combining visual data with other sensory inputs, is a growing trend in sports analytics. This approach enhances the robustness and depth of analysis by incorporating diverse data sources like GPS, biometric sensors, and audio inputs alongside video footage. For instance, combining video-based tracking with wearable GPS devices can provide comprehensive insights into player positioning and movement patterns. This multi-modal integration is important for creating more accurate and context-aware analytical models, as it allows for the synthesis of complementary data streams, offering a holistic view of player performance and game dynamics.

2.4.2 Potential Advancements and Their Impact on Sports Analytics

The future of sports analytics lies in the continuous advancement of computer vision technologies and their integration with other modalities. Potential advancements include the development of more sophisticated machine learning models that can handle the complexities of sports environments, such as rapid and unpredictable movements, occlusions, and varying lighting conditions. These models will likely employ more advanced neural network architectures, such as transformers and graph neural networks, to improve accuracy and efficiency.

However, it is important to recognize that the effectiveness of these methods is not solely dependent on model advancements. The availability of large-scale, high-quality annotated datasets is equally critical for driving performance improvements. Publicly sharing more data with high-quality annotations can significantly boost the performance of these models. To address the challenge of manual annotation, which can be resource-intensive, active learning techniques offer a promising solution. By selectively annotating the most informative samples, these techniques can reduce the annotation burden while still improving model performance (e.g., [54]).

Furthermore, the adoption of these technologies is expected to extend beyond elite sports, becoming accessible to amateur and youth sports through cost-effective solutions. For example, the study on action spotting transfer capabilities across diverse soccer domains [23], highlights the importance of these approaches in mitigating the differences between professional and amateur sports footage. By automating the extraction and analysis of meaningful information from video footage and leveraging these advancements in both data availability and machine learning, the sports analytics field will see more comprehensive and detailed insights into player performance, team strategies, and overall game dynamics. These advancements will ultimately enhance coaching, training, and broadcasting across various levels of sports, from elite to amateur and youth levels.

2.5 Summary

This chapter has provided a comprehensive overview of the critical elements of computer vision in sports analytics, including field registration, object tracking, Re-ID, action recognition, and pose estimation. These components collectively enable the automated extraction of meaningful insights from sports footage, advancing the analysis of player movements and team tactics. The importance of datasets in training and evaluating these computer vision models was also discussed, highlighting various benchmarks used in the sports domain. Additionally, the integration of multiple modalities and advanced machine learning techniques, such as deep learning and transformers, were explored to enhance the robustness and accuracy of these analytical models. These advancements are essential for achieving precise data acquisition and comprehensive analysis in sports.

Computer vision plays a pivotal role in advancing sports analytics by providing automated, detailed, and accurate insights into player performance and game dynamics. The technology reduces the reliance on manual data collection and labeling, making advanced analytics accessible to a broader range of sports, from elite to amateur levels. As the transition to the next chapter, which focuses on predictive analysis and play evaluation using machine learning, it is essential to acknowledge how the data obtained through computer vision serves as the foundation for these advanced analytical techniques. The integration of machine learning with rich visual data enables predictive models that can anticipate player actions and evaluate plays, further enhancing strategic decision-making in sports. This synergy between computer vision and machine learning marks a significant step forward in the evolution of sports analytics.

References

1. Abu Farha, Y., Richard, A., Gall, J.: When will you do what?-anticipating temporal occurrences of activities. In: Proceedings of the IEEE Conference on Computer Vision and Pattern Recognition, pp. 5343–5352 (2018)
2. Akan, S., Varlı, S.: Reidentifying soccer players in broadcast videos using body feature alignment based on pose. In: Proceedings of the 2023 4th International Conference on Computing, Networks and Internet of Things, pp. 440–444 (2023)
3. Akan, S., Varlı, S., Bhuiyan, M.A.N.: An enhanced swin transformer for soccer player reidentification. Sci. Rep. **14**(1), 1139 (2024)
4. Ali, M.M.N., Abdullah-Al-Wadud, M., Lee, S.-L.: An efficient algorithm for detection of soccer ball and players. In: Proceedings of 16th ASTL Control and Networking, pp. 39–46 (2012)
5. An, Q., Cui, K., Liu, R., Wang, C., Qi, M., Ma, H.: Attention-aware multiple granularities network for player re-identification. In: Proceedings of the 5th International ACM Workshop on Multimedia Content Analysis in Sports, pp. 137–144 (2022)
6. Andriluka, M., Pishchulin, L., Gehler, P., Schiele, B.: 2d human pose estimation: new benchmark and state of the art analysis. In: Proceedings of the IEEE Conference on computer Vision and Pattern Recognition, pp. 3686–3693 (2014)
7. Arbués-Sangüesa, A., Ballester, C., Haro, G.: Single-camera basketball tracker through pose and semantic feature fusion (2019). arXiv:1906.02042
8. Arbues-Sanguesa, A., Martín, A., Fernández, J., Ballester, C., Haro, G.: Using player's body-orientation to model pass feasibility in soccer. In: Proceedings of the IEEE/CVF Conference on Computer Vision and Pattern Recognition Workshops, pp. 886–887 (2020)
9. Askari, F., Ramaprasad, R., Clark, J.J., Levine, M.D.: Interaction classification with key actor detection in multi-person sports videos. In: Proceedings of the IEEE/CVF Conference on Computer Vision and Pattern Recognition, pp. 3580–3588 (2022)
10. Askari, F., Yared, C., Ramaprasad, R., Garg, D., Hu, A., Clark, J.J.: Video interaction recognition using an attention augmented relational network and skeleton data. In: Proceedings of the IEEE/CVF Conference on Computer Vision and Pattern Recognition, pp. 3225–3234 (2024)

11. Ballan, L., Bertini, M., Del Bimbo, A., Nunziati, W.: Soccer players identification based on visual local features. In: Proceedings of the 6th ACM International Conference on Image and Video Retrieval, pp. 258–265 (2007)

12. Bansal, S., Arora, C., Jawahar, C.V.: My view is the best view: procedure learning from egocentric videos. In: European Conference on Computer Vision, pp. 657–675. Springer (2022)

13. Baumgartner, T., Klatt, S.: Monocular 3d human pose estimation for sports broadcasts using partial sports field registration. In: Proceedings of the IEEE/CVF Conference on Computer Vision and Pattern Recognition, pp. 5109–5118 (2023)

14. Behrmann, N., Alireza Golestaneh, S., Kolter, Z., Gall, J., Noroozi, M.: Unified fully and timestamp supervised temporal action segmentation via sequence to sequence translation. In: European Conference on Computer Vision, pp. 52–68. Springer (2022)

15. Ben-Ami, I., Basha, T., Avidan, S.: Racing bib numbers recognition. In: Proceedings of the British Machine Vision Conference, pp. 1–10 (2012)

16. Benzakour, Y., Cabado, B., Giancola, S., Cioppa, A., Ghanem, B., Van Droogenbroeck, M.: Osl-actionspotting: a unified library for action spotting in sports videos. In: 2024 IEEE International Workshop on Sport, Technology and Research (STAR), pp. 132–137. IEEE (2024)

17. Bergmann, P., Meinhardt, T., Leal-Taixe, L.: Tracking without bells and whistles. In: Proceedings of the IEEE/CVF International Conference on Computer Vision, pp. 941–951 (2019)

18. Bewley, A., Ge, Z., Ott, L., Ramos, F., Upcroft, B.: Simple online and realtime tracking. In: 2016 IEEE International Conference on Image Processing (ICIP), pp. 3464–3468. IEEE (2016)

19. Biermann, H., Theiner, J., Bassek, M., Raabe, D., Memmert, D., Ewerth, R.: A unified taxonomy and multimodal dataset for events in invasion games. In: Proceedings of the 4th International Workshop on Multimedia Content Analysis in Sports, pp. 1–10 (2021)

20. Brasó, G., Leal-Taixé, L.: Learning a neural solver for multiple object tracking. In: Proceedings of the IEEE/CVF Conference on Computer Vision and Pattern Recognition, pp. 6247–6257 (2020)

21. Bridgeman, L., Volino, M., Guillemaut, J.-Y., Hilton, A.: Multi-person 3d pose estimation and tracking in sports. In: Proceedings of the IEEE/CVF Conference on Computer Vision and Pattern Recognition Workshops, pp. 0–0 (2019)

22. Bright, J., Balaji, B., Chen, Y., Clausi, D.A., Zelek, J.S.: Pitchernet: powering the moneyball evolution in baseball video analytics. In: Proceedings of the IEEE/CVF Conference on Computer Vision and Pattern Recognition, pp. 3420–3429 (2024)

23. Cabado, B., Cioppa, A., Giancola, S., Villa, A., Guijarro-Berdinas, B., Padrón, E.J., Ghanem, B., Van Droogenbroeck, M.: Beyond the premier: assessing action spotting transfer capability across diverse domains. In: Proceedings of the IEEE/CVF Conference on Computer Vision and Pattern Recognition, pp. 3386–3398 (2024)

24. Cao, J., Pang, J., Weng, X., Khirodkar, R., Kitani, K.: Observation-centric sort: Rethinking sort for robust multi-object tracking. In: Proceedings of the IEEE/CVF Conference on Computer Vision and Pattern Recognition, pp. 9686–9696 (2023)

25. Cao, Z., Hidalgo, G., Simon, T., Wei, S.-E., Sheikh, Y.: OpenPose: realtime multi-person 2D pose estimation using Part Affinity Fields (2018). arXiv:1812.08008

26. Carion, N., Massa, F., Synnaeve, G., Usunier, N., Kirillov, A., Zagoruyko, S.: End-to-end object detection with transformers. In: Proceedings of the 16th European Conference on Computer Vision, pp. 213–229. Springer (2020)

27. Carr, P., Sheikh, Y., Matthews, I.: Point-less calibration: camera parameters from gradient-based alignment to edge images. In: 2012 IEEE Workshop on the Applications of Computer Vision (WACV), pp. 377–384. IEEE (2012)

28. Chao, V., Jamsrandorj, A., Oo, Y.M., Mun, K.-R., Kim, J.: 3d ball trajectory reconstruction of a ballistic shot from a monocular basketball video. In: IECON 2023-49th Annual Conference of the IEEE Industrial Electronics Society, pp. 1–6. IEEE (2023)

29. Chen, J., Little, J.J.: Sports camera calibration via synthetic data. In: Proceedings of the IEEE/CVF Conference on Computer Vision and Pattern Recognition Workshops, pp. 0–0 (2019)
30. Chen, X., Tian, Y., Huang, Q., Zhang, W., Yu, Z.: Dynamic model based ball trajectory prediction for a robot ping-pong player. In: 2010 IEEE International Conference on Robotics and Biomimetics, pp. 603–608. IEEE (2010)
31. Cheng, B., Xiao, B., Wang, J., Shi, H., Huang, T.S., Zhang, L.: Higherhrnet: Scale-aware representation learning for bottom-up human pose estimation. In: Proceedings of the IEEE/CVF Conference on Computer Vision and Pattern Recognition, pp. 5386–5395 (2020)
32. Cheshire, E., Hu, M.-C., Chang, M.-H.: Player tracking and analysis of basketball plays. In: European Conference of Computer Vision (2015)
33. Chu, Y.-J., Su, J.-W., Hsiao, K.-W., Lien, C.-Y., Fan, S.-H., Hu, M.-C., Lee, R.-R., Yao, C.-Y., Chu, H.-K.: Sports field registration via keypoints-aware label condition. In: Proceedings of the IEEE/CVF Conference on Computer Vision and Pattern Recognition, pp. 3523–3530 (2022)
34. Cioppa, A., Deliege, A., Giancola, S., Ghanem, B., Van Droogenbroeck, M., Gade, R., Moeslund, T.B.: A context-aware loss function for action spotting in soccer videos. In: Proceedings of the IEEE/CVF Conference on Computer Vision and Pattern Recognition, pp. 13126–13136 (2020)
35. Cioppa, A., Deliege, A., Giancola, S., Ghanem, B., Van Droogenbroeck, M.: Scaling up soccernet with multi-view spatial localization and re-identification. Sci. Data **9**(1), 355 (2022)
36. Cioppa, A., Giancola, S., Deliege, A., Kang, L., Zhou, X., Cheng, Z., Ghanem, B., Van Droogenbroeck, M.: Soccernet-tracking: multiple object tracking dataset and benchmark in soccer videos. In: Proceedings of the IEEE/CVF Conference on Computer Vision and Pattern Recognition, pp. 3491–3502 (2022)
37. Cioppa, A., Giancola, S., Somers, V., Magera, F., Zhou, X., Mkhallati, H., Deliège, A., Held, J., Hinojosa, C., Mansourian, A.M., et al.: Soccernet 2023 challenges results. Sports Eng. **27**(2), 24 (2024)
38. Citraro, L., Márquez-Neila, P., Savare, S., Jayaram, V., Dubout, C., Renaut, F., Hasfura, A., Ben Shitrit, H., Fua, P.: Real-time camera pose estimation for sports fields. Mach. Vis. Appl. **31**(3), 16 (2020)
39. Claasen, P.J., De Villiers, J.P.: Video-based sequential bayesian homography estimation for soccer field registration. Exp. Syst. Appl. **252**, 124156 (2024)
40. Cui, Y., Zeng, C., Zhao, X., Yang, Y., Wu, G., Wang, L.: SportsMOT: a large multi-object tracking dataset in multiple sports scenes. In: Proceedings of the IEEE/CVF International Conference on Computer Vision, pp. 9921–9931 (2023)
41. De Vleeschouwer, C., Chen, F., Delannay, D., Parisot, C., Chaudy, C., Martrou, E., Cavallaro, A., et al.: Distributed video acquisition and annotation for sport-event summarization. New Eur. Media Summit **8** (2008)
42. Deliege, A., Cioppa, A., Giancola, S., Seikavandi, M.J., Dueholm, J.V., Nasrollahi, K., Ghanem, B., Moeslund, T.B., Van Droogenbroeck, M.: Soccernet-v2: a dataset and benchmarks for holistic understanding of broadcast soccer videos. In: 7th International Workshop on Computer Vision in Sports (CVsports) at IEEE/CVF Conference on Computer Vision and Pattern Recognition (CVPR'21), pp. 4508–4519 (2021)
43. Direkoglu, C., Sah, M., O'connor, N.: Player detection in field sports. Mach. Vis. Appl. **29**(2), 187–206 (2018)
44. D'Orazio, T., Leo, M., Mosca, N., Spagnolo, P., Mazzeo, P.L.: A semi-automatic system for ground truth generation of soccer video sequences. In: 2009 Sixth IEEE International Conference on Advanced Video and Signal Based Surveillance, pp. 559–564. IEEE (2009)
45. Duan, K., Bai, S., Xie, L., Qi, H., Huang, Q., Tian, Q.: Centernet: keypoint triplets for object detection. In: Proceedings of the IEEE/CVF International Conference on Computer Vision, pp. 6569–6578 (2019)
46. Farha, Y.A., Gall, J.: Ms-tcn: Multi-stage temporal convolutional network for action segmentation. In: Proceedings of the IEEE/CVF Conference on Computer Vision and Pattern Recognition, pp. 3575–3584 (2019)

47. Figueroa, P., Leite, N., Barros, R.M.L., Cohen, I., Medioni, G.: Tracking soccer players using the graph representation. In: Proceedings of the 17th International Conference on Pattern Recognition, pp. 787–790 (2004)

48. Fischler, M.A., Bolles, R.C.: Random sample consensus: a paradigm for model fitting with applications to image analysis and automated cartography. Commun. ACM **24**(6), 381–395 (1981)

49. Furuno, S., Kobayashi, K., Okubo, T., Kurihara, Y.: A study on spin-rate measurement using a uniquely marked moving ball. In: 2009 ICCAS-SICE, pp. 3439–3442. IEEE (2009)

50. Gan, Z., Jin, L., Cheng, Y., Cheng, Y., Teng, Y., Li, Z., Li, Y., Yang, W., Zhu, Z., Xing, J., et al.: Skatingverse: a large-scale benchmark for comprehensive evaluation on human action understanding. IET Comput, Vis (2024)

51. Gao, Y., Tebbe, J., Zell, A.: Robust stroke recognition via vision and imu in robotic table tennis. In: Artificial Neural Networks and Machine Learning–ICANN 2021: 30th International Conference on Artificial Neural Networks, Bratislava, Slovakia, September 14–17, 2021, Proceedings, Part I 30, pp. 379–390. Springer (2021)

52. Ghasemzadeh, S.A., Van Zandycke, G., Istasse, M., Sayez, N., Moshtaghpour, A., De Vleeschouwer, C.: Deepsportlab: a unified framework for ball detection, player instance segmentation and pose estimation in team sports scenes. In: The 32nd British Machine Vision Conference (2021)

53. Giancola, S., Amine, M., Dghaily, T., Ghanem, B.: Soccernet: a scalable dataset for action spotting in soccer videos. In: Proceedings of the IEEE Conference on Computer Vision and Pattern Recognition Workshops, pp. 1711–1721 (2018)

54. Giancola, S., Cioppa, A., Georgieva, J., Billingham, J., Serner, A., Peek, K., Ghanem, B., Van Droogenbroeck, M.: Towards active learning for action spotting in association football videos. In: Proceedings of the IEEE/CVF Conference on Computer Vision and Pattern Recognition, pp. 5098–5108 (2023)

55. Giancola, S., Ghanem, B.: Temporally-aware feature pooling for action spotting in soccer broadcasts. In: Proceedings of the IEEE/CVF Conference on Computer Vision and Pattern Recognition, pp. 4490–4499 (2021)

56. Glover, J., Kaelbling, L.P.: Tracking the spin on a ping pong ball with the quaternion bingham filter. In: 2014 IEEE International Conference on Robotics and Automation (ICRA), pp. 4133–4140. IEEE (2014)

57. Gossard, T., Krismer, J., Ziegler, A., Tebbe, J., Zell, A.: Table tennis ball spin estimation with an event camera. In: Proceedings of the IEEE/CVF Conference on Computer Vision and Pattern Recognition, pp. 3347–3356 (2024)

58. Gutiérrez-Pérez, M., Agudo, A.: No bells just whistles: sports field registration by leveraging geometric properties. In: Proceedings of the IEEE/CVF Conference on Computer Vision and Pattern Recognition, pp. 3325–3334 (2024)

59. Habel, K., Deuser, F., Oswald, N.: Clip-reident: contrastive training for player re-identification. In: Proceedings of the 5th International ACM Workshop on Multimedia Content Analysis in Sports, pp. 129–135 (2022)

60. Hartley, R., Zisserman, A.: Multiple View Geometry in Computer Vision. Cambridge University Press (2003)

61. Hayet, J.B., Mathes, T., Czyz, J., Piater, J., Verly, J., Macq, B.: A modular multi-camera framework for team sports tracking. In: IEEE Conference on Advanced Video and Signal Based Surveillance, pp. 493–498 (2005)

62. He, K., Zhang, X., Ren, S., Sun, J.: Deep residual learning for image recognition. In: Proceedings of the IEEE Conference on Computer Vision and Pattern Recognition, pp. 770–778 (2016)

63. He, S., Luo, H., Wang, P., Wang, F., Li, H., Jiang, W.: Transreid: transformer-based object re-identification. In: 2021 IEEE/CVF International Conference on Computer Vision (ICCV), pp. 14993–15002 (2021)

64. Held, J., Cioppa, A., Giancola, S., Hamdi, A., Ghanem, B., Van Droogenbroeck, M.: Vars: video assistant referee system for automated soccer decision making from multiple views. In:

Proceedings of the IEEE/CVF Conference on Computer Vision and Pattern Recognition, pp. 5086–5097 (2023)

65. Held, J., Itani, H., Cioppa, A., Giancola, S., Ghanem, B., Van Droogenbroeck, M.: X-vars: introducing explainability in football refereeing with multi-modal large language models. In: Proceedings of the IEEE/CVF Conference on Computer Vision and Pattern Recognition, pp. 3267–3279 (2024)

66. Herath, S., Harandi, M., Porikli, F.: Going deeper into action recognition: a survey. Image Vis. Comput. **60**, 4–21 (2017)

67. Homayounfar, N., Fidler, S., Urtasun, R.: Sports field localization via deep structured models. In: Proceedings of the IEEE Conference on Computer Vision and Pattern Recognition, pp. 5212–5220 (2017)

68. Honda, Y., Kawakami, R., Yoshihashi, R., Kato, K., Naemura, T.: Pass receiver prediction in soccer using video and players' trajectories. In: Proceedings of the IEEE/CVF Conference on Computer Vision and Pattern Recognition, pp. 3503–3512 (2022)

69. Hong, J., Zhang, H., Gharbi, M., Fisher, M., Fatahalian, K.: Spotting temporally precise, fine-grained events in video. In: European Conference on Computer Vision, pp. 33–51. Springer (2022)

70. Horiuchi, Y., Makino, Y., Shinoda, H.: Computational foresight: forecasting human body motion in real-time for reducing delays in interactive system. In: Proceedings of the 2017 ACM International Conference on Interactive Surfaces and Spaces, pp. 312–317 (2017)

71. Hu, Q., Scott, A., Yeung, C., Fujii, K.: Basketball-sort: an association method for complex multi-object occlusion problems in basketball multi-object tracking. Multimed, Tools Appl (2024)

72. Huang, B., Zhang, T., Wang, Y.: Pose2uv: single-shot multiperson mesh recovery with deep uv prior. IEEE Trans. Image Process. **31**, 4679–4692 (2022)

73. Huang, H.-W., Yang, C.-Y., Ramkumar, S., Huang, C.-I., Hwang, J.-N., Kim, P.-K., Lee, K., Kim, K.: Observation centric and central distance recovery for athlete tracking. In: Proceedings of the IEEE/CVF Winter Conference on Applications of Computer Vision, pp. 454–460 (2023)

74. Huang, H.-W., Yang, C.-Y., Sun, J., Kim, P.-K., Kim, K.-J., Lee, K., Huang, C.-I., Hwang, J.-N.: Iterative scale-up expansioniou and deep features association for multi-object tracking in sports. In: Proceedings of the IEEE/CVF Winter Conference on Applications of Computer Vision, pp. 163–172 (2024)

75. Ibrahim, M.S., Muralidharan, S., Deng, Z., Vahdat, A., Mori, G.: A hierarchical deep temporal model for group activity recognition. In: 2016 IEEE Conference on Computer Vision and Pattern Recognition (CVPR) (2016)

76. Ibrahim, M.S., Muralidharan, S., Deng, Z., Vahdat, A., Mori, G.: A hierarchical deep temporal model for group activity recognition. In: Proceedings of the IEEE Conference on Computer Vision and Pattern Recognition, pp. 1971–1980 (2016)

77. Ingwersen, C.K., Mikkelstrup, C., Jensen, J.N., Hannemose, M.R., Dahl, A.B.: Sportspose: a dynamic 3d sports pose dataset. In: Proceedings of the IEEE/CVF International Workshop on Computer Vision in Sports (2023)

78. Iskakov, K., Burkov, E., Lempitsky, V., Malkov, Y.: Learnable triangulation of human pose. In: Proceedings of the IEEE/CVF International Conference on Computer Vision, pp. 7718–7727 (2019)

79. Ivankovic, Z., Rackovic, M., Ivkovic, M.: Automatic player position detection in basketball games. Multimed. Tools Appl. **72**, 2741–2767 (2014)

80. Jiale, F., Calvin, Y., Fujii, K.: Foul prediction with estimated poses from soccer broadcast video (2024). arXiv:2402.09650

81. Jiang, W., Gamboa Higuera, J.C., Angles, B., Sun, W., Javan, M, Yi, K.M.: Optimizing through learned errors for accurate sports field registration. In: Proceedings of the IEEE/CVF Winter Conference on Applications of Computer Vision, pp. 201–210 (2020)

82. Jiang, Y., Cui, K., Chen, L., Wang, C., Xu, C.: Soccerdb: a large-scale database for comprehensive video understanding. In: Proceedings of the 3rd International Workshop on Multimedia Content Analysis in Sports, pp. 1–8 (2020)

83. Jin, S., Xu, L., Xu, J., Wang, C., Liu, W., Qian, C., Ouyang, W., Luo, P.: Whole-body human pose estimation in the wild. In: Computer Vision–ECCV 2020: 16th European Conference, Glasgow, UK, August 23–28, 2020, Proceedings, Part IX 16, pp. 196–214. Springer (2020)
84. Johnson, S., Everingham, M.: Clustered pose and nonlinear appearance models for human pose estimation. In: Procedings of the British Machine Vision Conference, pp. 1–11. Aberystwyth, UK (2010)
85. Kazemi, V., Burenius, M., Azizpour, H., Sullivan, J.: Multi-view body part recognition with random forests. In: 2013 24th British Machine Vision Conference. British Machine Vision Association (2013)
86. Koshkina, M., Elder, J.H.: A general framework for jersey number recognition in sports video. In: Proceedings of the IEEE/CVF Conference on Computer Vision and Pattern Recognition, pp. 3235–3244 (2024)
87. Koshkina, M., Pidaparthy, H., Elder, J.H.: Contrastive learning for sports video: unsupervised player classification. In: Proceedings of the IEEE/CVF Conference on Computer Vision and Pattern Recognition, pp. 4528–4536 (2021)
88. Kuehne, H., Arslan, A., Serre, T.: The language of actions: recovering the syntax and semantics of goal-directed human activities. In: Proceedings of the IEEE Conference on Computer Vision and Pattern Recognition, pp. 780–787 (2014)
89. Kurach, K., Raichuk, A., Stańczyk, P., Zając, M., Bachem, O., Espeholt, L., Riquelme, C., Vincent, D., Michalski, M., Bousquet, O., et al.: Google research football: a novel reinforcement learning environment. In: Proceedings of the AAAI Conference on Artificial Intelligence, vol. 34, pp. 4501–4510 (2020)
90. Lee, S.-E., Shibata, K., Nonaka, S., Nobuhara, S., Nishino, K.: Extrinsic camera calibration from a moving person. IEEE Robot. Autom. Lett. **7**(4), 10344–10351 (2022)
91. Lefevre, S., Fluck, C., Maillard, B., Vincent, N.: A fast snake-based method to track football player. In: International Conference on Machine Vision Applications (2000)
92. Li, H., Deng, A., Ke, Q., Liu, J., Rahmani, H., Guo, Y., Schiele, B., Chen, C.: Sports-QA: a large-scale video question answering benchmark for complex and professional sports (2024). arXiv:2401.01505
93. Li, R., Yang, S., Ross, D.A., Kanazawa, A.: Ai choreographer: music conditioned 3d dance generation with aist++. In: Proceedings of the IEEE/CVF International Conference on Computer Vision, pp. 13401–13412 (2021)
94. Li, S., Cao, Q., Liu, L., Yang, K., Liu, S., Hou, J., Yi, S.: Groupformer: group activity recognition with clustered spatial-temporal transformer. In: Proceedings of the IEEE/CVF International Conference on Computer Vision, pp. 13668–13677 (2021)
95. Li, W., Liu, H., Ding, R., Liu, M., Wang, P.: Lifting Transformer for 3d Human Pose Estimation in Video, vol. 2, p. 2 (2021). arXiv:2103.14304
96. Li, Y., Li, Y., Vasconcelos, N.: Resound: towards action recognition without representation bias. In: Proceedings of the European Conference on Computer Vision (ECCV), pp. 513–528 (2018)
97. Li, Y., Chen, L., He, R., Wang, Z., Wu, G., Wang, L.: Multisports: a multi-person video dataset of spatio-temporally localized sports actions. In: Proceedings of the IEEE/CVF International Conference on Computer Vision, pp. 13536–13545 (2021)
98. Li, Y.-H., Liu, K.-Y., Liu, S.-L., Feng, L., Qiao, H.: Involving distinguished temporal graph convolutional networks for skeleton-based temporal action segmentation. IEEE Trans. Circuits Syst. Video Technol. **34**(1), 647–660 (2023)
99. Lin, T.-Y., Goyal, P., Girshick, R., He, K., Dollár, P.: Focal loss for dense object detection. In: Proceedings of the IEEE International Conference on Computer Vision, pp. 2980–2988 (2017)
100. Liu, H., Chen, Q., Tan, Z., Liu, J.-J., Wang, J., Su, X., Li, X., Yao, K., Han, J., Ding, E., et al.: Group pose: a simple baseline for end-to-end multi-person pose estimation. In: Proceedings of the IEEE/CVF International Conference on Computer Vision, pp. 15029–15038 (2023)
101. Liu, P., Wang, J.-H.: Monotrack: shuttle trajectory reconstruction from monocular badminton video. In: Proceedings of the IEEE/CVF Conference on Computer Vision and Pattern Recognition, pp. 3513–3522 (2022)

102. Liu, S., Zhang, A., Li, Y., Zhou, J., Li, X., Dong, Z., Zhang, R.: Temporal segmentation of fine-gained semantic action: a motion-centered figure skating dataset. In: Proceedings of the AAAI Conference on Artificial Intelligence, vol. 35, pp. 2163–2171 (2021)
103. Liu, Y., Cheng, X., Ikenaga, T.: Motion-aware and data-independent model based multi-view 3d pose refinement for volleyball spike analysis. Multimed. Tools Appl. **83**(8), 22995–23018 (2024)
104. Lu, K., Chen, J., Little, J.J., He, H.: Light cascaded convolutional neural networks for accurate player detection. In: British Machine Vision Conference (BMVC) (2017)
105. Lu, Z., Elhamifar, E.: FACT: Frame-action cross-attention temporal modeling for efficient supervised action segmentation. In: Conference on Computer Vision and Pattern Recognition, vol. 2024, pp. 18175–18185 (2024)
106. Mackowiak, S., Konieczny, J., Kurc, M., Maćkowiak, P.: Football player detection in video broadcast. Comput. Vis. Graph. **6375**, 118–125 (2010). (Lecture Notes in Computer Science)
107. Magera, F., Hoyoux, T., Barnich, O., Van Droogenbroeck, M.: A universal protocol to benchmark camera calibration for sports. In: Proceedings of the IEEE/CVF Conference on Computer Vision and Pattern Recognition, pp. 3335–3346 (2024)
108. Maglo, A., Orcesi, A., Pham, Q.-C.: Efficient tracking of team sport players with few game-specific annotations. In: Proceedings of the IEEE/CVF Conference on Computer Vision and Pattern Recognition, pp. 3461–3471 (2022)
109. Majeed, F., Gilal, N.U., Al-Thelaya, K., Yang, Y., Agus, M., Schneider, J.: Mv-soccer: motion-vector augmented instance segmentation for soccer player tracking. In: Proceedings of the IEEE/CVF Conference on Computer Vision and Pattern Recognition, pp. 3245–3255 (2024)
110. Manafifard, M., Ebadi, H., Abrishami Moghaddam, H.: A survey on player tracking in soccer videos. Comput. Vis. Image Underst. **159**, 19–46 (2017)
111. Mansourian, A.M., Somers, V., De Vleeschouwer, C., Kasaei, S.: Multi-task learning for joint re-identification, team affiliation, and role classification for sports visual tracking. In: Proceedings of the 6th International Workshop on Multimedia Content Analysis in Sports, pp. 103–112 (2023)
112. Martinez, J., Hossain, R., Romero, J., Little, J.J.: A simple yet effective baseline for 3d human pose estimation. In: Proceedings of the IEEE International Conference on Computer Vision, pp. 2640–2649 (2017)
113. Mehraban, S., Adeli, V., Taati, B.: Motionagformer: enhancing 3d human pose estimation with a transformer-gcnformer network. In: Proceedings of the IEEE/CVF Winter Conference on Applications of Computer Vision, pp. 6920–6930 (2024)
114. Meinhardt, T., Kirillov, A., Leal-Taixe, L., Feichtenhofer, C.: Trackformer: multi-object tracking with transformers. In: Proceedings of the IEEE/CVF Conference on Computer Vision and Pattern Recognition, pp. 8844–8854 (2022)
115. Metzler, J., Pagel, F.: 3d trajectory reconstruction of the soccer ball for single static camera systems. In: IAPR International Conference on Pattern Recognition, pp. 121–124 (2013)
116. Mkhallati, H., Cioppa, A., Giancola, S., Ghanem, B., Van Droogenbroeck, M.: Soccernet-caption: dense video captioning for soccer broadcasts commentaries. In: Proceedings of the IEEE/CVF Conference on Computer Vision and Pattern Recognition, pp. 5074–5085 (2023)
117. Naidoo, W.C., Tapamo, J.R.: Soccer video analysis by ball, player and referee tracking. In: Proceedings of the Annual Research Conference of the South African Institute of Computer Scientists and Information Technologists (SAICSIT) on IT Research in Developing Countries, pp. 51–60 (2006)
118. Nakabayashi, T., Higa, K., Yamaguchi, M., Fujiwara, R., Saito, H.: Event-based ball spin estimation in sports. In: Proceedings of the IEEE/CVF Conference on Computer Vision and Pattern Recognition, pp. 3367–3375 (2024)
119. Nakatani, C., Kawashima, H., Ukita, N.: Learning group activity features through person attribute prediction. In: Proceedings of the IEEE/CVF Conference on Computer Vision and Pattern Recognition, pp. 18233–18242 (2024)
120. Napolean, Y., Wibowo, P.T., van Gemert, J.C.: Running event visualization using videos from multiple cameras. In: Proceedings Proceedings of the 2nd International Workshop on Multimedia Content Analysis in Sports, pp. 82–90 (2019)

121. Nibali, A., Millward, J., He, Z., Morgan, S.: ASPset: an outdoor sports pose video dataset with 3D keypoint annotations. In: Image and Vision Computing, pp. 104196 (2021)
122. Nie, X., Chen, S., Hamid, R.: A robust and efficient framework for sports-field registration. In: Proceedings of the IEEE/CVF Winter Conference on Applications of Computer Vision, pp. 1936–1944 (2021)
123. Nonaka, N., Fujihira, R., Koshiba, T., Maeda, A., Seita, J.: Rugby scene classification enhanced by vision language model. In: Proceedings of the IEEE/CVF Conference on Computer Vision and Pattern Recognition, pp. 3256–3266 (2024)
124. Nonaka, N., Fujihira, R., Nishio, M., Murakami, H., Tajima, T., Yamada, M., Maeda, A., Seita, J.: End-to-end high-risk tackle detection system for rugby. In: Proceedings of the IEEE/CVF Conference on Computer Vision and Pattern Recognition (CVPR) Workshops, pp. 3550–3559 (June 2022)
125. Okamoto, L., Parmar, P.: Hierarchical neurosymbolic approach for comprehensive and explainable action quality assessment. In: Proceedings of the IEEE/CVF Conference on Computer Vision and Pattern Recognition, pp. 3204–3213 (2024)
126. Pan, J.-H., Gao, J., Zheng, W.-S.: Action assessment by joint relation graphs. In: Proceedings of the IEEE/CVF International Conference on Computer Vision, pp. 6331–6340 (2019)
127. Parmar, P., Morris, B.: Action quality assessment across multiple actions. In: 2019 IEEE Winter Conference on Applications of Computer Vision (WACV), pp. 1468–1476. IEEE (2019)
128. Parmar, P., Morris, B.T.: Learning to score olympic events. In: Proceedings of the IEEE Conference on Computer Vision and Pattern Recognition Workshops, pp. 20–28 (2017)
129. Pei Xu, K., Yang, Y., Xu, Y.: Person re-identification with end-to-end scene text recognition. In: Computer Vision: Second CCF Chinese Conference, CCCV 2017, Tianjin, China, October 11–14, 2017, Proceedings, Part III, pp. 363–374. Springer (2017)
130. Penate-Sanchez, A., Freire-Obregon, D., Lorenzo-Melian, A., Lorenzo-Navarro, J., Castrillon-Santana, M.: TGC20ReId: a dataset for sport event re-identification in the wild. Pattern Recogn. Lett. **138**, 355–361 (2020)
131. Pettersen, S.A., Johansen, D., Johansen, H., Berg-Johansen, V., Gaddam, V.R., Mortensen, A., Langseth, R., Griwodz, C., Stensland, H.K., Halvorsen, P.: Soccer video and player position dataset. In: Proceedings of the 5th ACM Multimedia Systems Conference, pp. 18–23 (2014)
132. Pishchulin, L., Insafutdinov, E., Tang, S., Andres, B., Andriluka, M., Gehler, P.V., Schiele, B.: Deepcut: joint subset partition and labeling for multi person pose estimation. In: Proceedings of the IEEE Conference on Computer Vision and Pattern Recognition, pp. 4929–4937 (2016)
133. Rahimian, P., Toka, L.: Optical tracking in team sports: a survey on player and ball tracking methods in soccer and other team sports. J. Quant. Anal. Sports **18**(1), 35–57 (2022)
134. Rasouli, A., Kotseruba, I., Tsotsos, J.K.: Are they going to cross? A benchmark dataset and baseline for pedestrian crosswalk behavior. In: Proceedings of the IEEE International Conference on Computer Vision Workshops, pp. 206–213 (2017)
135. Reddy, N.D., Guigues, L., Pishchulin, L., Eledath, J., Narasimhan, S.G.: Tessetrack: end-to-end learnable multi-person articulated 3d pose tracking. In: Proceedings of the IEEE/CVF Conference on Computer Vision and Pattern Recognition, pp. 15190–15200 (2021)
136. Redmon, J., Divvala, S., Girshick, R., Farhadi, A.: You only look once: unified, real-time object detection. In: Proceedings of the IEEE Conference on Computer Vision and Pattern Recognition, pp. 779–788 (2016)
137. Rhodin, H., Spörri, J., Katircioglu, I., Constantin, V., Meyer, F., Müller, E., Salzmann, M., Fua, P.: Learning monocular 3d human pose estimation from multi-view images. In: Proceedings of the IEEE Conference on Computer Vision and Pattern Recognition, pp. 8437–8446 (2018)
138. Rodriguez, M.D., Ahmed, J., Shah, M.: Action mach a spatio-temporal maximum average correlation height filter for action recognition. In: 2008 IEEE Conference on Computer Vision and Pattern Recognition, pp. 1–8. IEEE (2008)
139. Rodriguez-Criado, D., Bachiller-Burgos, P., Vogiatzis, G., Manso, L.J.: Multi-person 3d pose estimation from unlabelled data. Mach. Vis. Appl. **35**(3), 1–18 (2024)

140. Sato, S., Aono, M.: Leveraging human pose estimation model for stroke classification in table tennis. In: In Working Notes Proceedings of the MediaEval 2020 Workshop (2020)
141. Scott, A., Uchida, I., Ding, N., Umemoto, R., Bunker, R., Kobayashi, R., Koyama, T., Onishi, M., Kameda, Y., Fujii, K.: Teamtrack: a dataset for multi-sport multi-object tracking in full-pitch videos. In: Proceedings of the IEEE/CVF Conference on Computer Vision and Pattern Recognition, pp. 3357–3366 (2024)
142. Scott, A., Uchida, I., Onishi, M., Kameda, Y., Fukui, K., Fujii, K.: Soccertrack: a dataset and tracking algorithm for soccer with fish-eye and drone videos. In: 8th International Workshop on Computer Vision in Sports (CVsports) at IEEE/CVF Conference on Computer Vision and Pattern Recognition (CVPR' 22), pp. 3569–3579 (2022)
143. Senocak, A., Oh, T.-H., Kim, J., Kweon, I.S.: Part-based player identification using deep convolutional representation and multi-scale pooling. In: Proceedings of the IEEE Conference on Computer Vision and Pattern Recognition Workshops, pp. 1732–1739 (2018)
144. Seweryn, K., Wróblewska, A., Łukasik, S.: Survey of action recognition, spotting and spatio-temporal localization in soccer–current trends and research perspectives (2023). arXiv:2309.12067
145. Sha, L., Hobbs, J., Felsen, P., Wei, X., Lucey, P., Ganguly, S.: End-to-end camera calibration for broadcast videos. In: Proceedings of the IEEE/CVF Conference on Computer Vision and Pattern Recognition, pp. 13627–13636 (2020)
146. Shao, D., Zhao, Y., Dai, B., Lin, D.: Finegym: a hierarchical video dataset for fine-grained action understanding. In: Proceedings of the IEEE/CVF Conference on Computer Vision and Pattern Recognition, pp. 2616–2625 (2020)
147. Sharma, R.A., Bhat, B., Gandhi, V., Jawahar, C.V.: Automated top view registration of broadcast football videos. In: 2018 IEEE Winter Conference on Applications of Computer Vision (WACV), pp. 305–313. IEEE (2018)
148. Shen, Y., Wang, L., Jin, Y.: Aaformer: a multi-modal transformer network for aerial agricultural images. In: Proceedings of the IEEE/CVF Conference on Computer Vision and Pattern Recognition (CVPR) Workshops, pp. 1705–1711 (June 2022)
149. Shi, D., Wei, X., Li, L., Ren, Y., Tan, W.: End-to-end multi-person pose estimation with transformers. In: Proceedings of the IEEE/CVF Conference on Computer Vision and Pattern Recognition, pp. 11069–11078 (2022)
150. Somers, V., Joos, V., Cioppa, A., Giancola, S., Abolfazl Ghasemzadeh, S., Magera, F., Standaert, B., Mansourian, A.M., Zhou, X., Kasaei, S., et al.: Soccernet game state reconstruction: end-to-end athlete tracking and identification on a minimap. In: Proceedings of the IEEE/CVF Conference on Computer Vision and Pattern Recognition, pp. 3293–3305 (2024)
151. Suda, S., Makino, Y., Shinoda, H.: Prediction of volleyball trajectory using skeletal motions of setter player. In: Proceedings of the 10th Augmented Human International Conference 2019, pp. 1–8 (2019)
152. Suglia, A., Lopes, J., Bastianelli, E., Vanzo, A., Agarwal, S., Nikandrou, M., Yu, L., Konstas, I., Rieser, V.: Going for goal: a resource for grounded football commentaries (2022). arXiv:2211.04534
153. Sun, K., Xiao, B., Liu, D., Wang, J.: Deep high-resolution representation learning for human pose estimation. In: Proceedings of the IEEE/CVF Conference on Computer Vision and Pattern Recognition, pp. 5693–5703 (2019)
154. Sun, P., Cao, J., Jiang, Y., Yuan, Z., Bai, S., Kitani, K., Luo, P.: Dancetrack: multi-object tracking in uniform appearance and diverse motion. In: Proceedings of the IEEE/CVF Conference on Computer Vision and Pattern Recognition, pp. 20993–21002 (2022)
155. Sun, P., Cao, J., Jiang, Y., Zhang, R., Xie, E., Yuan, Z., Wang, C., Luo, P.: Transtrack: multiple object tracking with transformer (2020). arXiv:2012.15460
156. Suzuki, T., Takeda, K., Fujii, K.: Automatic fault detection in race walking from a smartphone camera via fine-tuning pose estimation. In: 2022 IEEE 11th Global Conference on Consumer Electronics (GCCE), pp. 631–632. IEEE (2022)
157. Suzuki, T., Takeda, K., Fujii, K.: Automatic detection of faults in simulated race walking from a fixed smartphone camera. Int. J. Comput. Sci. Sport **23**(1), 22–36 (2024)

158. Suzuki, T., Tanaka, R., Takeda, K., Fujii, K.: Pseudo-label based unsupervised fine-tuning of a monocular 3d pose estimation model for sports motions. In: Proceedings of the IEEE/CVF Conference on Computer Vision and Pattern Recognition, pp. 3315–3324 (2024)

159. Suzuki, T., Tsutsui, K., Takeda, K., Fujii, K.: Runner re-identification from single-view running video in the open-world setting. Multimed. Tools Appl. 1–17 (2024)

160. Tan, C., Sun, T., Fu, T., Wang, Y., Xu, M., Liu, S.: Hierarchical spatial-temporal network for skeleton-based temporal action segmentation. In: Chinese Conference on Pattern Recognition and Computer Vision (PRCV), pp. 28–39. Springer (2023)

161. Tanaka, R., Suzuki, T., Fujii, K.: 3d pose-based temporal action segmentation for figure skating: a fine-grained and jump procedure-aware annotation approach. In: Proceedings of the 7th International Workshop on Multimedia Content Analysis in Sports (2024)

162. Tanaka, R., Suzuki, T., Takeda, K., Fujii, K.: Automatic edge error judgment in figure skating using 3d pose estimation from a monocular camera and imus. In: Proceedings of the 6th International Workshop on Multimedia Content Analysis in Sports, pp. 41–48 (2023)

163. Tarashima, S.: Sflnet: direct sports field localization via cnn-based regression. In: Pattern Recognition: 5th Asian Conference, ACPR 2019, Auckland, New Zealand, November 26–29, 2019, Revised Selected Papers, Part I, vol. 5, pp. 677–690. Springer (2020)

164. Tekin, B., Rozantsev, A., Lepetit, V., Fua, P.: Direct prediction of 3d body poses from motion compensated sequences. In: Proceedings of the IEEE Conference on Computer Vision and Pattern Recognition, pp. 991–1000 (2016)

165. Theiner, J., Ewerth, R.: Tvcalib: camera calibration for sports field registration in soccer. In: Proceedings of the IEEE/CVF Winter Conference on Applications of Computer Vision, pp. 1166–1175 (2023)

166. Thomas, G., Gade, R., Moeslund, T.B., Carr, P., Hilton, A.: Computer vision for sports:current applications and research topics. Comput. Vis. Image Underst. **159**, 3–18 (2017)

167. Toshev, A., Szegedy, C.: Deeppose: human pose estimation via deep neural networks. In: Proceedings of the IEEE Conference on Computer Vision and Pattern Recognition, pp. 1653–1660 (2014)

168. Uchida, I., Scott, A., Shishido, H., Kameda, Y.: Automated offside detection by Spatio-Temporal analysis of football videos. In: Proceedings of the 4th International Workshop on Multimedia Content Analysis in Sports. Association for Computing Machinery, pp. 17–24 (2021)

169. Van Zandycke, G, De Vleeschouwer, C.: 3d ball localization from a single calibrated image. In: Proceedings of the IEEE/CVF Conference on Computer Vision and Pattern Recognition, pp. 3472–3480 (2022)

170. Van Zandycke, G., Somers, V., Istasse, M., Del Don, C., Zambrano, D.: Deepsportradar-v1: computer vision dataset for sports understanding with high quality annotations. In: Proceedings of the 5th International ACM Workshop on Multimedia Content Analysis in Sports, pp. 1–8 (2022)

171. Vats, K., McNally, W., Walters, P., Clausi, D.A., Zelek, J.S.: Ice hockey player identification via transformers and weakly supervised learning. In: Proceedings of the IEEE/CVF Conference on Computer Vision and Pattern Recognition, pp. 3451–3460 (2022)

172. Vats, K., Walters, P., Fani, M., Clausi, D.A., Zelek, J.S.: Player tracking and identification in ice hockey. Exp. Syst. Appl. **213**, 119250 (2023)

173. Wang, Y., Tran, D., Liao, Z.: Learning hierarchical poselets for human parsing. In: CVPR 2011, pp. 1705–1712. IEEE (2011)

174. Wojke, N., Bewley, A., Paulus, D.: Simple online and realtime tracking with a deep association metric. In: 2017 IEEE International Conference on Image Processing (ICIP), pp. 3645–3649. IEEE (2017)

175. Wu, L.-F., Wang, Q., Jian, M., Qiao, Y., Zhao, B.-X.: A comprehensive review of group activity recognition in videos. Int. J. Autom. Comput. **18**(3), 334–350 (2021)

176. Wu, T., He, R., Wu, G., Wang, L.: Sportshhi: a dataset for human-human interaction detection in sports videos. In: Proceedings of the IEEE/CVF Conference on Computer Vision and Pattern Recognition, pp. 18537–18546 (2024)

177. Xarles, A., Escalera, S., Moeslund, T.B., Clapés, A.: T-deed: temporal-discriminability enhancer encoder-decoder for precise event spotting in sports videos. In: Proceedings of the IEEE/CVF Conference on Computer Vision and Pattern Recognition, pp. 3410–3419 (2024)

178. Xu, C., Fu, Y., Bing, Z., Zitian, C., Yu-Gang, J., Xiangyang, X.: Learning to score figure skating sport videos. IEEE Trans. Circuits Syst. Video Technol. **30**(12), 4578–4590 (2019)

179. Xu, J., Rao, Y., Yu, X., Chen, G., Zhou, J., Lu, J.: Finediving: a fine-grained dataset for procedure-aware action quality assessment. In: Proceedings of the IEEE/CVF Conference on Computer Vision and Pattern Recognition, pp. 2949–2958 (2022)

180. Xu, Y., Osep, A., Ban, Y., Horaud, R., Leal-Taixé, L., Alameda-Pineda, X.: How to train your deep multi-object tracker. In: Proceedings of the IEEE/CVF Conference on Computer Vision and Pattern Recognition, pp. 6787–6796 (2020)

181. Xu, Y., Jing, Z., Qiming, Z., Dacheng, T.: Vitpose: simple vision transformer baselines for human pose estimation. Adv. Neural Inf. Process. Syst. **35**, 38571–38584 (2022)

182. Yan, R., Xie, L., Tang, J., Shu, X., Tian, Q.: Social adaptive module for weakly-supervised group activity recognition. In: Proceedings of the 16th European Conference on Computer Vision, pp. 208–224. Springer (2020)

183. Yang, F., Odashima, S., Masui, S., Jiang, S.: Hard to track objects with irregular motions and similar appearances? Make it easier by buffering the matching space. In: Proceedings of the IEEE/CVF Winter Conference on Applications of Computer Vision, pp. 4799–4808 (2023)

184. Ye, M., Shen, J., Lin, G., Xiang, T., Shao, L., Hoi, S.C.H.: Deep learning for person re-identification: a survey and outlook. IEEE Trans. Pattern Anal. Mach. Intell. **44**(6), 2872–2893 (2021)

185. Yeung, C., Ide, K., Fujii, K.: Autosoccerpose: automated 3d posture analysis of soccer shot movements. In: Proceedings of the IEEE/CVF Conference on Computer Vision and Pattern Recognition, pp. 3214–3224 (2024)

186. Yi, F., Wen, H., Jiang, T.: Asformer: transformer for action segmentation. In: The British Machine Vision Conference (BMVC), pp. 1–15 (2021)

187. Zalluhoglu, C., Ikizler-Cinbis, N.: Collective sports: a multi-task dataset for collective activity recognition. Image Vis. Comput. **94**, 103870 (2020)

188. Zhang, N., Izquierdo, E.: A high accuracy camera calibration method for sport videos. In: 2021 International Conference on Visual Communications and Image Processing (VCIP), pp. 1–5. IEEE (2021)

189. Zhang, N., Izquierdo, E.: A four-point camera calibration method for sport videos. IEEE Trans. Circuits Syst, Video Technol (2023)

190. Zhang, S.-H., Li, R., Dong, X., Rosin, P., Cai, Z., Han, X., Yang, D., Huang, H., Hu, S.-M.: Pose2seg: detection free human instance segmentation. In: Proceedings of the IEEE/CVF Conference on Computer Vision and Pattern Recognition, pp. 889–898 (2019)

191. Zhang, Y., Sun, P., Jiang, Y., Yu, D., Weng, F., Yuan, Z., Luo, P., Liu, W., Wang, X.: Byte-track: multi-object tracking by associating every detection box. In: European Conference on Computer Vision, pp. 1–21. Springer (2022)

192. Zhao, Y., Li, Z., Chen, K.: A method for tracking hockey players by exploiting multiple detections and omni-scale appearance features. Proj, Rep (2020)

193. Zhong, Y., Demiris, Y.: Dancemvp: self-supervised learning for multi-task primitive-based dance performance assessment via transformer text prompting. In Proceedings of the AAAI Conference on Artificial Intelligence **38**, 10270–10278 (2024)

194. Zhou, K., Yang, Y., Cavallaro, A., Xiang, T.: Omni-scale feature learning for person re-identification. In: Proceedings of the IEEE/CVF International Conference on Computer Vision, pp. 3702–3712 (2019)

195. Zhou, X., Kang, L., Cheng, Z., He, B., Xin, J.: Feature combination meets attention: Baidu soccer embeddings and transformer based temporal detection (2021). arXiv:2106.14447

Chapter 3
Predictive Analysis and Play Evaluation with Machine Learning

Abstract This chapter examines the important role of machine learning in sports predictive analysis and play evaluation. It covers a spectrum of techniques, from traditional result analysis to advanced machine learning approaches, addressing key areas such as game result, event, and trajectory prediction, as well as action and space evaluation in team sports. The chapter introduces various datasets and methodologies, highlighting the evolution from rule-based systems to deep learning models. It explores how these techniques are applied to classify plays, cluster similar behaviors, extract meaningful features, and learn complex representations from sports data. The discussion extends to counterfactual analysis, providing insights into hypothetical scenarios and their potential impacts. By presenting cutting-edge research and future directions, including posture analysis and real-time analytics, this chapter offers a comprehensive view of how data-driven approaches advance sports analytics, enhancing our understanding of individual and team performances, and informing tactic decision-making in sports.

Keywords Machine learning · Predictive analysis · Play evaluation · Space evaluation · Counterfactual analysis

3.1 Introduction

Predictive analysis has become a vital component in sports, providing valuable insights that influence both individual and tactical decisions. By leveraging machine learning, analysts can evaluate player performance, predict game outcomes, and understand complex team tactics. The advancements in machine learning, including neural network approaches, have significantly enhanced our ability to process and interpret large datasets, offering a more detailed understanding of sports behaviors and outcomes. This chapter explores the impact of these technologies on predictive analysis and play evaluation, highlighting their applications and potential in various sports.

K. Fujii, *Machine Learning in Sports*,
SpringerBriefs in Computer Science, https://doi.org/10.1007/978-981-96-1445-5_3

This chapter introduces the application of machine learning in sports predictive analysis and play evaluation, focusing on several key areas. Section 3.2 first introduces the dataset we can use in as well as in Chaps. 1 and 2. Section 3.3 describes traditional result data analysis is introduced to understand its history, limitations, and challenges. Next, the latter of this section introduces machine learning applications for sports behavior analysis, covering classification and clustering methods, and highlighting recent approaches to feature extraction and representation learning. The chapter then moves on to predictive analysis and play evaluations in Sect. 3.4, discussing methods for outcome prediction, event prediction, space evaluation, and trajectory prediction in team sports. Finally, Sects. 3.5 and 3.6 introduce counterfactual analysis techniques and discuss future topics, including posture analysis and other applications, setting the stage for further advancements in sports analytics. These sections show a comprehensive overview of how machine learning is advancing the analysis and evaluation of individual performance and tactics.

3.2   Datasets for Predictive Analysis and Play Evaluation

In Chap. 2, datasets accompanied by video footage are introduced, which provide rich visual context for analysis. This chapter mainly focuses on event and tracking data, which offer detailed information about specific actions and player movements usually during more games than video datasets. These datasets are crucial for understanding the movements and interactions within the game. However, note that pose data, which provides precise body position and movement details, is not publicly available without accompanying video footage.

In soccer, several comprehensive datasets are available for event data as described in Chaps. 1 and 2. Wyscout, in collaboration with Pappalardo et al. [108], provides open spatiotemporal event data for matches from the top five European leagues, EURO 2016, and the 2018 World Cup. Statsbomb also offers open event data, including lineup and match metadata of the top five European leagues and international championships such as EURO 2020, 2022, and 2024, and the 2022 World Cup, accessible via their GitHub repository.[1] Statsbomb also provides 360 freeze frame data, which includes the positions of all players at the moment of key events captured in broadcast footage, for the above international championships. For tracking data, Metrica Sports offers tracking data and events for top-tier football matches,[2] and the data in three matches are publicly available. Opta, a division of Stats Perform, is well-known for its rich event and tracking data and is widely used by analysts and clubs globally, whereas DataStadium Inc. also provides high-quality tracking and event data in Japan, but both are not publicly available. Tracking data is generally paid and/or restricted access, and its limited availability poses challenges to the democratization of sports analytics. For example, knowledge about various

---

[1] https://github.com/statsbomb/open-data.

[2] https://github.com/metrica-sports/sample-data.

pre-processing (e.g., synchronizing tracking data with event data [140]) and main processing techniques is not widely shared.

In basketball, tracking data have recently been analyzed, but data sources are limited. Most studies have relied on the only open-source dataset[3] from the National Basketball Association (NBA) 2015–2016 season, pre-processed by the STATS SportVU system, now known as STATS Perform. This dataset includes the trajectories of basketball players and the ball at 25 frames per second from approximately 630 games. The positional data contains the XY coordinates of each player on the court and the XYZ coordinates of the ball. Even in 2024, this dataset remains widely used because no alternative open-source dataset has emerged. Additionally, the dataset's size is sufficient for most machine learning models. However, as time passes, this dataset will become outdated and be necessary for more large-scale trajectory datasets in the future. Although not considering trajectory data, there have been studies analyzing games using movement annotations, such as those in the FIBA Basketball Champions League [128] and the Liga ACB Spanish Championship [76]. In other invasion team sports, handball datasets provide video, tracking, and event data in [4, 98] and American football tracking and event data in NFL are available.[4] Tracking data for other invasion team sports, such as rugby, as well as field and ice hockey, may not be publicly available, although there might be some datasets that exist but are not widely known.

3.3 Result Analysis in Match and Play

Historically, observable outcomes such as the number of shots and passes have been manually counted and analyzed through visual observation. However, this approach only handles events that can be consistently defined by any observer and requires the same amount of time to reanalyze as it did initially (e.g., [57]). This makes it difficult to quantify the frequency of common patterns in sports, such as "various potential next plays that experienced players can narrow down based on their knowledge", or "plays that are defined differently by experienced players but are important for understanding the game" (for example, see [151]). Evaluating such plays is naturally challenging as well.

On the other hand, by using machine learning techniques to process player and ball positions along with event data, it is possible to include expert knowledge and analyze plays based on a transparent set of criteria. Although it might not be entirely persuasive to many individuals, this approach allows for repeated recalculations and the processing of large datasets, offering significant advantages over visual observation.

---

[3] https://github.com/rajshah4/BasketballData.

[4] https://github.com/asonty/ngs_highlights.

3.3.1 Traditional Approach

In traditional methods that do not utilize a learning-based approach, researchers have relied on their experience and established theories to evaluate the characteristics of multi-agent behaviors. For example, based on hypotheses, they have calculated the distances and relative phases of two athletes (e.g., [5, 44, 137]), the speeds of movements (e.g., [121]), and the frequencies and angles of actions such as shots and passes (e.g., [21, 45, 49, 143]). These analyses often involve the computation of representative values such as averages and maximums. Measurement systems with high spatiotemporal resolution, like motion capture systems and force platforms, have been used to analyze skillful maneuvers in terms of cognition, force, and torque (e.g., [38, 43]). After obtaining these representative values, specific hypotheses are tested, sometimes using statistical analysis (e.g., [44, 111]).

For instance, in basketball, defensive cooperation against team attacks, known as screen-plays, which block the movements of a defender, has been evaluated. The results showed that defenders flexibly change their roles (switching, overlapping, ignoring, or providing global help) according to the level of urgency [44]. This traditional quantitative approach is powerful, applicable to small datasets, and easy to interpret across various fields because it allows for the direct testing of hypotheses. In soccer, for example, after estimating ball trajectory from videos, six typical soccer attack patterns for tactic analysis are defined in a rule-based manner [104]. Mathematical approaches have also been used to compute representative values. For example, players' areas of control in soccer have been represented as Voronoi diagrams [132] (for the details, see Sect. 3.4.3). Other studies have analyzed pass connections using network theory [156], self-similarity in team positioning [69], and the breaking of spatiotemporal symmetry using group theory in ball possession scenarios [162]. However, to represent cooperative and competitive interactions more flexibly and practically, advanced modeling techniques are needed.

3.3.2 Machine Learning Approaches for Result Analysis

To integrate machine learning into result analysis in sports analytics, feature extraction, and representation learning are crucial. Machine learning techniques process raw motion data to extract features, which are then used for tasks such as classification and regression, including play recognition and motion prediction. Here examples of learning-based classification in soccer and basketball analytics are mainly introduced.

One of the simplest applications of machine learning involves using static features and annotated labels in classification models. Examples include team identification in soccer [81], classification of ball possession [93] and attacking processes [70] in soccer, screen-play classification in basketball [57, 88, 89, 168], and predicting who will obtain a rebound in basketball [58], utilizing techniques such as linear

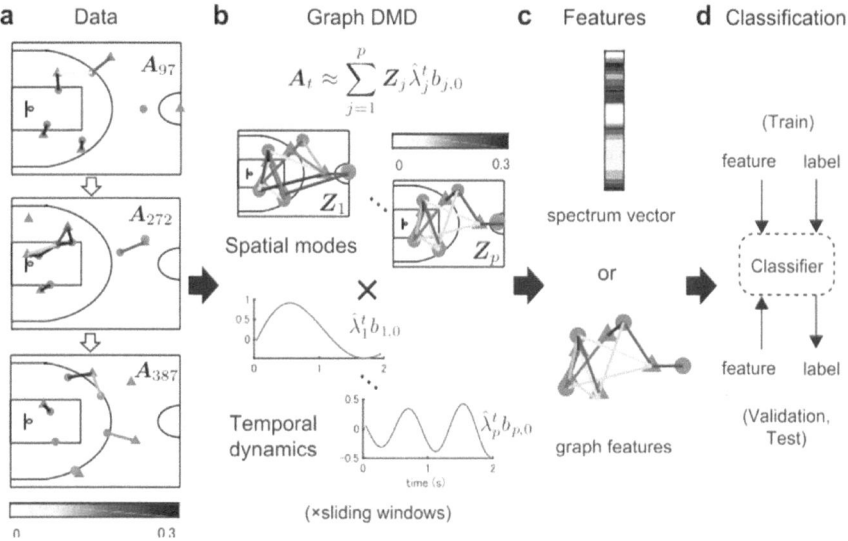

Fig. 3.1 Schematic diagram of our Graph DMD for team play classification [40]. **a** Input data: the grayscale spectrum in each edge represents the values as a function of the distance between agents; they approach 0 if far and 1 if near. The input of Graph DMD is the adjacency matrix series $\mathbf{A_t}$, where t is each timestamp. **b** Graph DMD decomposes the input into the sum of the product of graph (spatial) j-th modes $\mathbf{Z_j}$ and temporal dynamics λ_j (b represents an initial value). To analyze the time-varying dynamics, Graph DMD is performed for each temporal sliding window. **c** After Graph DMD, features for classification are computed in different approaches. One is simply to vectorize the Graph DMD modes (i.e., spectrum). The second is to compute graph features using existing methods. **d** Using feature vectors and labels, classification models for screen-play and zone defense are trained, validated, and tested. The figure is used in [40]

discriminant analysis (LDA), logistic regression, or support vector machine (SVM) with handcrafted static features like distances between players and angles, such as player-goal vectors on the court.

However, when applied to complex multi-agent movements, it is often necessary to account for spatiotemporal structures. A straightforward approach is to use spatiotemporal features derived from unsupervised learning techniques. Common methods include unsupervised learning techniques for spatial or multivariate inputs, such as principal component analysis (PCA), which aids in dimensionality reduction. Dimensionality reduction transforms high-dimensional data into meaningful lower-dimensional representations. For instance, PCA, factor analysis [50, 75], t-distributed stochastic neighbor embedding (t-SNE) [139] regarding shot types [85], non-negative matrix factorization (NMF) [94], tensor decomposition [106] regarding shot types, and topic modeling [95, 146] of trajectories have been used to summarize diverse interactive sports behaviors into lower-dimensional representations. However, some of these methods assume independence of sampling, meaning the extracted information does not reflect temporal properties.

Several approaches are used to reduce the number of dimensions while considering the time-series structures. For example, image-based approaches transform trajectory data into images using neural networks (e.g., [103, 145]), including the self-organizing map (e.g., [68]). Another approach for extracting physically interpretable dynamical properties is a method called *dynamic mode decomposition* (DMD), which is applied to screen-play and zone defense classifications [40] as illustrated in Fig. 3.1.

Clustering is an unsupervised learning technique that groups a set of objects into clusters, where objects within the same cluster exhibit higher similarity based on selected features compared to those in different clusters. This method is particularly useful for identifying natural groupings in data without predefined labels. In team sports data, researchers have utilized hierarchical clustering (e.g., [56]) based on similarity measures [24, 66, 125] and distribution-based clustering using Gaussian mixture models [110]. However, challenges arise with time-series data, as it is difficult to compute similarity when data lengths are not fixed. To address this, specific designs for time series similarity enable the application of conventional clustering methods. Hierarchical clustering requires appropriate distance measures, such as the Fréchet distance and dynamic time warping (DTW), used in basketball [16, 24, 125] and soccer [24]. Fast and scalable methods for computing Fréchet distance for trajectory mining [66, 131] and discriminative sub-trajectory mining [8] have also been developed.

Another challenge is calculating the distance or similarity between multi-agent trajectories. Simple methods for comparing agent-to-agent trajectories face permutation problems due to the constant swapping of players or roles in multiple sequences of plays [125]. This misalignment is inherent in raw multi-agent data. A rule-based method addresses this by permuting the players closest to the ball (e.g., [39, 40]). Alternatively, data-driven permutation techniques, such as the Hungarian algorithm (a linear assignment method), have been applied to role assignment problems in basketball, identifying roles like guard, forward, and center [125, 157]. Another way to handle the permutation problem is by calculating the similarity of multivariate nonlinear dynamical systems using DMD, as used in [37, 39]. Since DMD is a dimensionality reduction technique similar to PCA, it extracts dynamical properties that are invariant to permutations.

As machine learning evolves, neural networks have become instrumental in automating feature extraction. Various neural network architectures, such as recurrent neural networks (RNN) [145], temporal convolutional networks [103], graph neural networks (GNN) [2], and transformers [120], are considered for feature extraction. In particular, GNN and transformer approaches can learn permutation-equivariant features, which solve the permutation problem described above. Additionally, neural network approaches that compute trajectory similarities in scalable ways using a metric learning framework with a Siamese network have been proposed [149]. Other researchers have developed an interpretable classification model with attention mechanism [166]. These methods are highly scalable, suitable for large datasets, and capable of capturing abstract or complex representations. However, they require substantial amounts of data for training. Supervised learning provides a potent strategy when a large amount of labeled data is available, yet the associated disadvantages include big labor costs and often limited size of labeled datasets.

Semi-supervised learning approaches are one of the solutions, in which the model is trained on a dataset that contains labeled data and (usually much larger) unlabeled data. In team sports, semi-supervised methods have been applied to the extraction of tactical patterns in soccer [2, 31]. Self-supervised learning also utilizes unlabeled data by deriving supervisory signals through diverse pre-processing techniques. For example, a transformer-based approach was applied in basketball for recognizing group behaviors [54] and learning trajectory representations [1, 147]. Although semi- and self-supervised learning approaches are currently rare in team sports, the potential may grow in the future, particularly when we can access larger amounts of (unlabeled) data.

3.4 Predictive Analysis and Play Evaluations

This section introduces approaches in predictive analysis and play evaluations in four areas: game results prediction, event prediction and evaluation, space evaluation, and trajectory prediction. It explains how advanced models and data-driven techniques can predict the outcomes of team sports games, assess and forecast in-game events, evaluate the spatial aspects of play, and predict player and ball trajectories. Each of these subfields contributes to a comprehensive understanding of the game, providing valuable insights for coaches, analysts, and other stakeholders (e.g., [22, 134]).

3.4.1 Game Result Prediction

Predicting the outcomes of sports matches is a significant and complex task, especially in soccer, due to factors like low scoring and frequent draws. This complexity attracts various stakeholders, including fans, bookmakers, and team analysts. Accurate game result predictions are crucial for betting markets, while coaches and analysts use models to identify key match features to enhance team performance [158]. Notable efforts in soccer prediction include the 2017 and 2023 Soccer Prediction Challenges, which utilized the Open International Soccer Database containing over 216,000 matches from 52 leagues across 35 countries. These competitions have set benchmarks for evaluating predictive models, particularly emphasizing the Ranked Probability Score (RPS) as a key evaluation metric [19, 30].

The primary approaches employed for soccer match result prediction include statistical models, machine learning approaches, and hybrid techniques that combine elements of both. Statistical models often involve the use of distributions like the Poisson model to predict goal counts (e.g., [29, 83]), while ML models leverage algorithms such as gradient-boosted trees (e.g., CatBoost) [115] and deep learning frameworks [116]. Hybrid models, which integrate statistical approaches with machine learning techniques, have shown promise in improving prediction accuracy (e.g., [52, 72]). For instance, studies have demonstrated the effectiveness of combining random forest models with Poisson-based ranking systems (e.g., [52]). Additionally, to enhance prediction performance, ensemble methods like XGBoost and CatBoost have been applied successfully to soccer-specific ratings, such as pi-ratings [20] and Berrar ratings [3]. These methods have been further refined through the integration of spatiotemporal data and advanced feature selection techniques [67, 108], aiming to balance accuracy with model interpretability for practical applications in sports analytics.

3.4.2 Event Prediction and Evaluation

Play evaluation methods include machine learning, mathematical models, and their combinations. In team sports tactics, even plays with high success probabilities can be countered by opposing strategies. Consequently, there is no definitive single measure for player and team evaluations, then these assessments should be used as indicative references rather than absolute truths. This section introduces event prediction and evaluation, including scoring and other actions such as passes and interceptions.

The most popular example of event prediction is a scoring prediction, which has been investigated such as in basketball [12, 14, 37, 39, 80]. Based on a similar idea, the expected possession value [12, 13] have been developed to evaluate expected score values for each possession or attack, which was extended in soccer [35] and handball [73, 97]. When analyzing player compatibility in basketball (more scoring opportunities and substitutions), the performance of a particular lineup has been evaluated based on scoring efficiency as a regression analysis (called a lineup analysis). For example, player positions (roles) were redefined by clustering players based on play frequency and success rates to evaluate lineup efficiency [65]. Another example is Bayesian modeling by including a flag indicating the presence of two specific players on offense to estimate their impact on scoring efficiency [62]. Building on these studies, Yamada and Fujii exclusively focused on offensive features to quantitatively assess player compatibility [155] to provide deeper insights into lineup efficiency. For more comprehensive and interpretable analysis, to obtain interpretable spatial representations, researchers have developed several approaches to predict action from player locations, such as using matrix [163] and tensor [109] factor models, and Poisson point process model [96].

In soccer, having rare scoring opportunities and fewer substitutions, players or plays have been evaluated based on the expected values of goals [23, 86, 87, 107] (reviewed in [142]). In machine learning approaches, large amounts of data have been used to predict the probability of scoring or the probability of an on-ball event occurring (e.g., dribbles and passes), by evaluating actions as a "move that is likely to score points" [22]. The strength of this learning-based approach is that it can be modeled flexibly without using much sports knowledge.

However, evaluations based on score predictions are unstable (particularly, in soccer) because they predict events that are rare in the entire game, and it is difficult to evaluate the variety of plays leading up to points scored and lost. For this reason, methods have been proposed to evaluate a team's defense using player behavior and position information of all players and the ball, based on predictions of ball capture and effective attacks [135, 138] as shown in Fig. 3.2. This approach evaluates performance based not just on the results of actions (such as shots or goals) but on the process and decision-making leading up to those actions. For instance, a scenario where a shot results in a goal might not necessarily be evaluated as a poor defense if this probabilistic assessment of their actions is realized. Regarding passing actions, similarly, researchers have considered modeling and valuing of a pass [6, 48, 112–114, 118], pass-receiving [26, 45, 79], the defender's pass interception [117], pressing [92], and bulid-up defense [141].

In American football, a study examined the relationships between player actions and game outcome, highlighting how specific player behaviors and decisions impact the overall success of the team [18]. Using Bayesian non-parametric models, the expected hypothetical completion probability framework was developed to evaluate if quarterbacks throw passes to the receiver most likely to catch it [25]. Using a

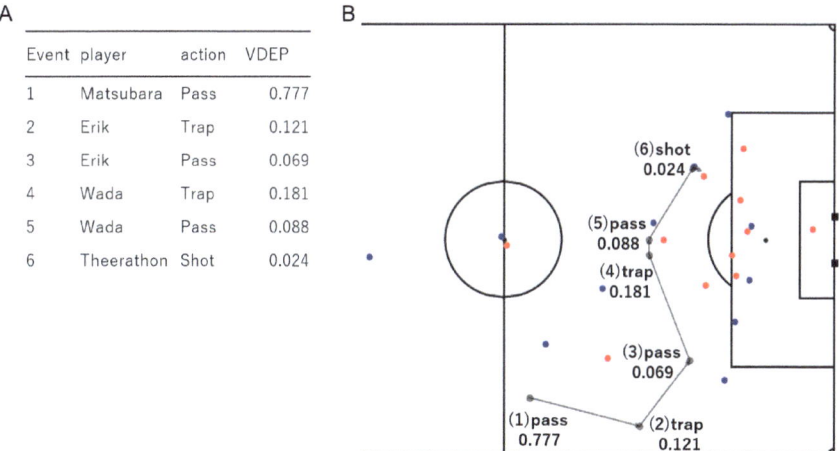

Fig. 3.2 Example of our defensive play analysis based on the event prediction via machine learning [135] called VDEP (Valuing Defense by Estimating Probabilities of ball recovering and being attacked). **a** The VDEP value for each event is indicated, including the type of event, and the player who took the action. Our classifiers for VDEP predict a ball recovery or an effective attack in the subsequent 5 actions or events. **b** The position of all players when the shot was performed is visualized (red: defending team; blue: attacking team) and the flow of the event with the ball. In this scene, the VDEP values were positive in all events, suggesting that the defensive performance was not poor enough to result in conceding a goal. The figure is used in [135]

neural network approach, within-play valuation models were proposed using player-tracking data to estimate the expected yards gained by the ball-carrier based on the positions and trajectories of all players on the field [164].

Transformer-based action prediction models have been also used for forecasting and evaluating player actions. Seq2Event model [126] predicts the next events and positions based on past events and context, and developed a practical metric, called possession utilization. Yeung et al. [161] introduced a transformer-based neural point process model to predict the next events, positions, and inter-event time. They also proposed the possession utilization score (HPUS) metric and examined its relationship with team performance indicators over a season, such as final rankings, average goals scored, and expected goals (xG), highlighting its utility in comprehensive football analysis. Recently, event modeling inspired by large language models has been proposed [90, 91].

3.4.3 Space and Off-Ball Player Evaluation

Although on-ball player performance is generally evaluated through event prediction, the movements of off-ball players are often not defined as events. Hence, off-ball player movements are frequently assessed using space evaluation metrics. Defensive play evaluation is essentially the counterpart of offense evaluation, focusing largely on on-ball events such as defensive pressure and ball recoveries, which can be evaluated based on event prediction in the previous subsection. As reviewed by [36], group-level defensive behavior and synchronization have been also investigated in soccer.

Rule-based evaluations of off-ball movements have been considered in basketball. For example, off-ball cutting was evaluated [130] and other researchers [76] show that a player's off-ball movements can substantially influence the probability of getting an open shot. An extensive model for evaluating off-ball movements in passing situations was also introduced [153], which also accounts for player profiles. In soccer, rule-based space evaluation about dangerousity has been considered [78]. Researchers have tried to evaluate off-ball players, but analyzing their spatial movements is highly complex due to the multiple potential actions in various situations.

Space evaluation has been also considered in the context of off-ball play evaluations. As mathematical models for space evaluation, the dominant area from the perspective of the Voronoi diagram [46, 132] was considered based on the minimum arrival time to determine which player can reach each area on the field most quickly, which sometimes takes into account velocity and acceleration. Such studies have extended to the estimation of the kinematic model for each player [7, 84] and weighting the field with the arrival times of the players [102]. More advanced studies have quantified off-ball scoring opportunities (OBSO) with probabilistic mathematical models [127] and proposed an index to evaluate the movement to create space by separating defensive players from other attacking players during a certain period [34]. Since OBSO model [127] is a versatile mathematical model, and in our group, it has been extended with the improved scoring model with defenders [134, 159] and extended to defenders in soccer [138], attackers in basketball [74] and Ultimate [64] by modification based on the sport specificity. In particular, our group [74] proposed two mathematical models to predict off-ball scoring opportunities in basketball, considering pass-to-score, dribble-to-score, and interception situations, called the Ball Intercept and Movement for Off-ball Scoring (BIMOS) models. In contrast, purely data-driven space evaluation is also possible, e.g., developed with a neural network approach in badminton doubles [28].

3.4.4 Trajectory Prediction

Although event prediction primarily focuses on forecasting specific events at given time points, machine learning approaches can also predict player trajectories. In the 1990s and 2000s, research on pedestrian prediction problems, mainly mathematical or rule-based models (see, e.g., [55]), was widely investigated. However, trajectory prediction of players in team sports, which involve complex movements, has been significantly accelerated by the advent of deep learning. For multi-agent trajectory prediction in sports such as basketball and soccer, many methods have employed RNNs [63, 77, 123, 167], including variational RNNs [41, 157, 165]. Additionally, some approaches have utilized generative adversarial networks (GANs) [15, 60], variational autoencoders [33], and transformer-based models [1, 11]. While most of these methods treat trajectory prediction as a straightforward forecasting task, a few studies have framed it as an imitation learning problem within the reinforcement learning framework, utilizing expert demonstrations [41, 77].

In many of these models, agents are assumed to have complete visibility of other agents to facilitate long-term centralized prediction [157, 165]. This assumption, however, often results in a lack of interpretability regarding the agents' internal states, such as which information each agent utilizes. To address this, attention-based observation techniques have been proposed for multi-agent systems in real-world applications [41, 53, 59]. Other methods, like neural relational inference, focus on estimating movement coupling in physical and biological systems [51, 71]. For more precise modeling, decentralized approaches [41] are necessary to compute individual agent observations and contributions accurately.

Recent studies have introduced new frameworks to enhance the accuracy and efficiency of multi-agent trajectory prediction. For example, "Team Game" approach that explicitly models interactions at multiple levels was proposed [150]. Semi-supervised generative models for multiagent trajectories were developed [32], while the focus on modeling conditional dependencies in multiagent trajectories using autoregressive models and GNN was addressed [119]. The challenge of off-screen behavior prediction in football was tackled with GNN [105] and diffusion models [61], further expanding the application of trajectory prediction in sports analytics. These advancements, combined with the use of permutation-equivariant features in graph neural networks [51, 71, 129, 157] and multi-head attention mechanisms in transformer [1], have significantly improved the ability to model complex multi-agent dynamics and predict trajectories more accurately.

Another important approach is to incorporate tactical perspectives into trajectory prediction. For example, trajectory predictions reflecting defensive evaluations in soccer [133], trajectory computation optimizing defensive evaluations in basketball [124], and offensive evaluations in soccer [27], have been developed. In American football, player trajectories by predicting league average movements [9, 122] were simulated and evaluated. In particular, research on trajectory prediction and space evaluation using mathematical models have developed separately, with the former alone being unable to evaluate players, and the latter alone being able to evaluate only the player receiving the ball. In this respect, the work of [134] in our group can evaluate sacrificed movements for teammates (for example, movements to create space) by quantifying every off-ball player's impact on scores in terms of the difference between predicted and real player trajectory, as illustrated in Fig. 3.3. In general, evaluating plays using counterfactual prediction is one of the important applications of trajectory prediction, which is explained in the next section.

3.5 Counterfactual Analysis

Counterfactual analysis in sports analytics offers a substantial utility that allows analysts to explore hypothetical scenarios and their potential impacts on game outcomes. This technique helps us understand the effects of different strategies, and player movements by considering various "what-if" situations. There are several approaches to conducting counterfactual analysis, each with its unique strengths and applications. These include conditioning within machine learning, causal inference, mathematical modeling, and simulation-based methods.

3.5.1 Conditioning in Machine Learning

In sports analytics, conditioning within machine learning involves training models on specific scenarios to evaluate what might happen if those situations change. For example, a model could be trained on a dataset to predict player movements based on the positions of their teammates and opponents. By conditioning on certain variables, such as the position of a key player with RNN [41, 157] and diffusion models [17] in basketball, the model can generate predictions for different hypothetical scenarios. This approach helps us understand how specific factors influence player behavior, providing insights into strategic decision-making.

In soccer, our group introduced the Shooting Payoff Computation (SPC) framework [159], which uses game theory and machine learning to analyze shot-taking scenarios in football. The framework includes the Expected Probability of Shot On

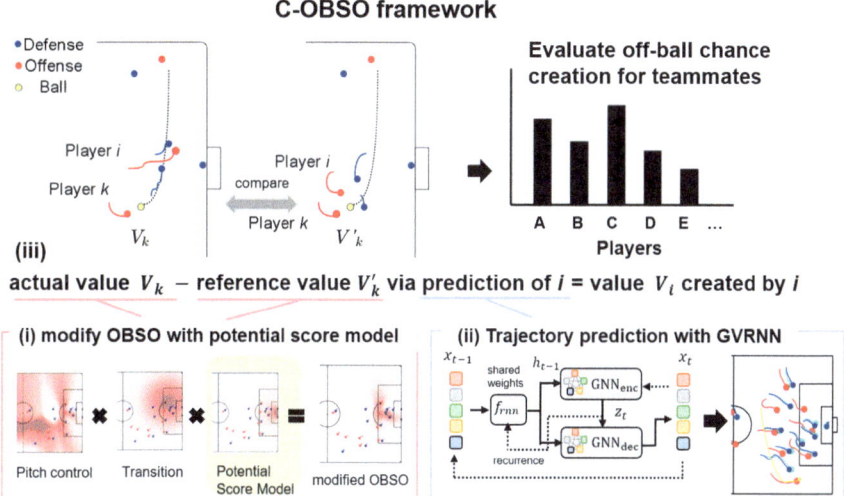

Fig. 3.3 Overview of our C-OBSO approach [134] (Creating Off-ball Scoring Opportunity). (i) First, a score model in OBSO [127] (it consists of a pitch control model to evaluate spatial control and area dominance on the soccer pitch, a transition model to predict the ball position in the next event, and score model to represent scoring probability when a shot is performed in the area) is modified as a potential score model by including defenders' positions. (ii) Then players' trajectories are predicted using Graph Variational RNN (GVRNN) [157] to generate a reference player trajectory. Each agent trajectory is processed via an RNN with shared parameters. The graph encoder and decoder model the relationship between agents, and finally output the movement of each agent in the next timestamp. (iii) Finally, a C-OBSO metric for player i is calculated by the difference between the evaluation value in the actual game V_k for future on-ball player k and the referenced or predicted value V_k' via GVRNN. From the arXiv version of [134], the figure is licensed under CC-BY-SA 4.0

Target (xSOT) metric based on deep learning, allowing for counterfactual analysis of shot situations, including determining optimal passing decisions. By developing the xSOT metric based on deep learning, the framework effectively evaluates player actions beyond just goals, allowing for detailed analysis and comparison of different shots and providing insights into optimal strategies as illustrated in Fig. 3.4.

In another soccer example, the TacticAI framework [148], developed with Liverpool FC experts, analyzes and optimizes player positioning during corner kicks in football. Incorporating predictive and generative components allows for counterfactual analysis, determining the most effective player setups. Validated through benchmark tasks and expert evaluations, TacticAI provides suggestions that are preferred over existing tactics 90% of the time, demonstrating its utility in improving tactical decisions. However, it is often limited by the bias of the training data (in particular when the training data is limited) and the inability to capture true causal effects beyond observed correlations.

Shooter: Emerson Palmieri dos Santos
Match: Italy vs. Wales (EURO 2020)

Jersey Number	P(S_{on})	P(S_{off})	P(S_{block})	P(Control)
9	0.27	0.32	0.22	0.59
20	0.23	0.60	0.03	0.63
14	0.17	0.67	0.16	0.99
12	0.15	0.63	0.20	0.89
6	0.05	0.53	0.18	0.17
Shooter	0.03	0.51	0.46	-
8	0.00	0.61	0.40	0.60

Fig. 3.4 Shot-taking situation example in our shooting payoff computation framework [159]. In the left figure, an example scene in the match of Italy vs. Wales in EURO 2020 is shown. In the right table, $P(S_{on})$ represents the probability of a shot being on target. $P(S_{off})$ is the probability of a shot being off target. $P(S_{block})$ indicates the probability of a shot being blocked by the defender. $P(S_{on})$, $P(S_{off})$, and $P(S_{block})$ are probabilities related to shot outcomes when the player is in possession of the ball. $P(Control)$ represents the probability of the attacker maintaining control of the ball when receiving a pass. Combined with the left figure, Attacker 9, positioned closest to the goal, showed the highest likelihood of an on-target shot (0.27) and the lowest chance of shooting off-target (0.32). Attacker 20, while further from the goal, had the second-best on-target probability (0.23) and faced the lowest risk of shot blockage (0.03), likely due to fewer defenders in their path. Attacker 14 demonstrated the highest ball control probability (0.99), presumably due to a lack of nearby defenders. Given these probabilities, passing to Attacker 14 appears to be the most strategic decision for maintaining possession and potentially creating better scoring opportunities because their high $P(Control)$ increases the likelihood of sustained possession. The figure and table are modified from [159]

3.5.2 Causal Inference

Causal inference in sports analytics is one of the statistical techniques to estimate the causal relationships between different factors and outcomes. This approach often involves the use of counterfactuals, where the goal is to understand what would have happened if a different action had been taken in the context of sports analytics. If we can perform randomized experiments by setting the same situations, it would be easy to estimate the effect of actions by comparing the outcomes with and without the actions, but in many sports situations, there is no practical or ethical way to conduct such randomized experiments. Instead, we must rely on observational data and advanced statistical methods to infer causal relationships and estimate the impact of various actions and strategies.

For example, in the static setting of causal inference for team sports, propensity score matching to reduce selection bias by matching units with similar propensity scores (i.e., the probability of receiving an intervention in given state; more precisely, within the framework of causal inference, it is the probability of a treatment assignment conditional on covariates) was used to investigate the causal effect of going for the touchdown in American football [154], clearing the puck in ice hockey [136], the effectiveness of timeouts in basketball [47], at stopping an opposing run crossing the ball in soccer [152], and baseball pitching [100].

In the dynamic setting, a g-computation method was applied [144] to examine the effect of a specific pitch (taking a pitch during a 3-0) in baseball. However, these approaches cannot estimate the effect of actions in a spatiotemporal multi-agent setting. Recently our work [42] has addressed this issue by counterfactual recurrent networks in multi-agent systems to estimate the effect of extra pass actions in a shot situation as illustrated in Fig. 3.5. This approach utilizes graph variational RNNs and domain-specific theory-based computation for individual treatment effect (ITE) estimation based on long-term predictions of multi-agent covariates and outcomes. These methods have performed realistic counterfactual predictions and evaluated the counterfactual passes in shot scenarios.

Outside the context of causal inference based on the Rubin causal model, the causality of the physics-based system such as using neural relational inference (e.g., [71]) has been considered (sometimes called causal reasoning). While more challenging to implement, causal methods can provide more robust and generalizable insights about the effects of actions in complex sports systems. If we know the mathematical structures of sports, we can utilize mathematical models for counterfactual prediction.

3.5.3 Mathematical Models

Mathematical models in counterfactual analysis involve creating mathematical representations of sports that can simulate different scenarios. Although causal inference in the previous subsection is a methodology developed with a statistical foundation that estimates underlying relationships from data. In contrast, mathematical models differ in that they regard the models of underlying relationships as predefined. The mathematical models are often based on well-established physical or mathematical rules. Examples of the mathematical models are already introduced in Sect. 3.4.3. By adjusting the parameters of the model, analysts can explore how changes in one aspect of the game might affect the overall outcome. This approach provides a structured and quantitative way to explore "what if" questions in sports. For example, Umemoto and Fujii [138] proposed an evaluation method for team soccer defense positioning by computing counterfactuals based on the OBSO model [127], identifying optimal defensive positions.

3.5.4 Rule-Based Simulation Models

Simulation-based approaches or rule-based models involve creating detailed models of the sport and simulating various scenarios to observe potential outcomes often in the longer term than mathematical models. This approach is particularly useful

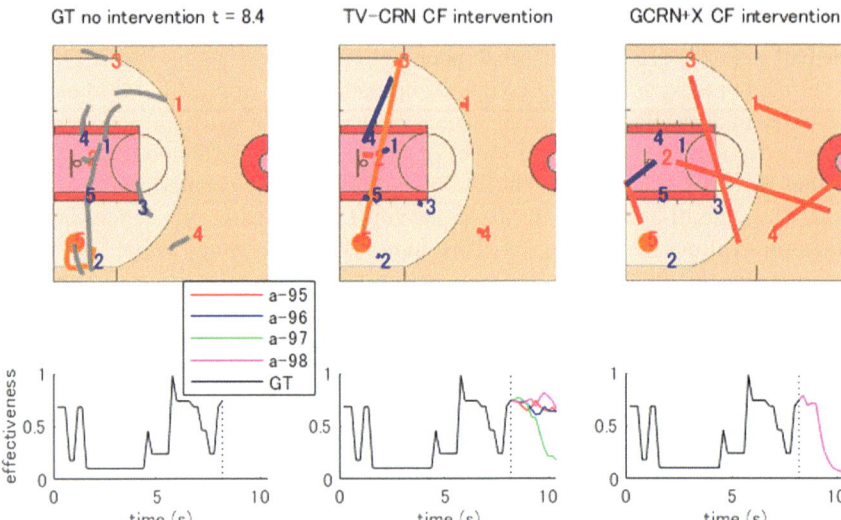

Fig. 3.5 Example of NBA counterfacutual analysis using our causal model approach [42]. Visualization of (top) trajectories and (bottom) outcome time series in (left) ground truth without intervention, (middle) counterfactual intervention using our model (TV-CRN: Theory-based variational counterfactual recurrent network), and (right) the baseline (modification of GCRN (Graph counterfactual recurrent network) [82] to predict the trajectory). Here the intervention is an extra pass in the shot situation. (Top) Red and blue numbers, gray line, and orange circle and line indicate an attacker, a defender, players' historical trajectories, and the ball, respectively. The positions of the numbers are at the end of the factual data (shot), which is shown as the break line in lower plots. In the CF (counterfactual) intervention subplots, colored trajectories indicate counterfactual predictions. The actual red player #5 shot (left) but in the counterfactual prediction (middle and right columns), the player tried to pass to a teammate. In the middle top (our method), the player successfully passes to the teammate red #3, but in the right top (baseline), the player's pass failed. (Bottom) outcome time series (attack effectiveness) are shown. We define the attack effectiveness as the outcome by predicting whether the attack is effective (defined in a rule-based manner [42]) or not at the next time stamp using logistic regression. The "a" in the lower caption is the intervention times. For example, "$a = 95$" means the case of intervention at the 95th frame (9.5 s). The figure is reprinted with permission from IEEE about the article [42]

for exploring complex interactions and dynamics that are difficult to capture with traditional statistical models. Although rule-based modeling can closely align with human intentions, it remains challenging to represent all events and perform counterfactual predictions in sports. For example, our group [99] proposes a new method for estimating the effect of various batting strategies in baseball using counterfactual simulations and a deep learning model that transforms batting ability based on strategy changes. The method found that employing different strategies can increase runs when switching costs are ignored, and it provides insights into the conditions under which multiple strategies are effective, offering a clearer understanding of their impact on game outcomes. In invasion team sports such as soccer and basketball, it would be necessary to consider movement-based agent models. The agent modeling is described in Chap. 4.

3.6 Future Research Topics

As future research topics, we can consider conducting next play analyses, such as offensive tactics, defensive maneuvers, or other player movements in the next game, from the perspective of several critical factors, including model validation, posture analysis, and real-time data processing. Evaluating the validity of play evaluation models is extremely challenging due to the lack of ground truth. Quantitative assessment can be conducted using event success probabilities or selection probabilities (such as in [148]), but these only measure model accuracy and are merely necessary conditions for validating the models. From the perspectives of team or player evaluations, it is possible to calculate correlations with existing metrics such as goals, player ratings in a game, and salaries (e.g., [101, 134]). However, since the goal is to assess elements not covered by these metrics, perfect alignment is neither achievable nor desirable. Thus, qualitative evaluation is also essential (e.g., [10]), but reliance on it should be minimized due to its subjectivity, particularly regarding tactical evaluation. In summary, it is essential to combine quantitative and qualitative evaluations to achieve a comprehensive assessment.

Pose information is crucial for play evaluation as it provides insights into a player's orientation and readiness for the next move. However, in team sports, while there are some datasets available for posture (see Chap. 2), this information is not yet widely linked to large-scale event data and tracking data, making it challenging to use for tactical analysis. Starting with focused analyses on specific scenarios like shooting, as demonstrated by [160], is a practical approach. Moving forward, the release of such comprehensive datasets will be essential for advancing tactical analysis.

Many of the computations introduced in this discussion, especially those involving deep learning, require a certain amount of time for training; however, the inference time is quite rapid. This means that once the data is collected at the end of a game, evaluations can be performed almost immediately. The primary bottleneck, therefore, lies in the acquisition of data from video and sensors. Automated processes for field detection, tracking, and Re-ID, although advanced, are not perfect and still require manual corrections. If complete data is not required, it is technically feasible to output evaluations in near real-time. Advancements in computer vision are expected to reduce processing time.

3.7 Summary

This chapter has provided a comprehensive overview of machine learning applications in sports predictive analysis and play evaluation. Various approaches are introduced, from traditional result analysis to advanced machine learning techniques, covering areas such as game result prediction, event prediction and evaluation, space evaluation, and trajectory prediction. The introduction of counterfactual analysis methods has shown how we can explore hypothetical scenarios to gain deeper insights into game dynamics. As the field continues to evolve, the integration of these techniques promises to provide increasingly sophisticated and actionable insights for coaches, analysts, and other stakeholders in the world of sports.

References

1. Alcorn, M.A., Nguyen, A.: baller2vec: a multi-entity transformer for multi-agent spatiotemporal modeling (2021). arXiv:2102.03291
2. Anzer, G., Bauer, P., Brefeld, U., Faßmeyer, D.: Detection of tactical patterns using semi-supervised graph neural networks. In: MIT Sloan Sports Analytics Conference, vol. 16, pp. 1–3 (2022)
3. Berrar, D., Lopes, P., Dubitzky, W.: Incorporating domain knowledge in machine learning for soccer outcome prediction. Mach. Learn. **108**, 97–126 (2019)
4. Biermann, H., Theiner, J., Bassek, M., Raabe, D., Memmert, D., Ewerth, R.: A unified taxonomy and multimodal dataset for events in invasion games. In: Proceedings of the 4th International Workshop on Multimedia Content Analysis in Sports, pp. 1–10 (2021)
5. Bourbousson, J., Sève, C., McGarry, T.: Space–time coordination dynamics in basketball: Part 2. The interaction between the two teams. J. Sports Sci. **28**(3), 349–358 (2010)
6. Bransen, L., Van Haaren, J., van de Velden, M.: Measuring soccer players' contributions to chance creation by valuing their passes. J. Quant. Anal. Sports **15**(2), 97–116 (2019)
7. Brefeld, U., Lasek, J., Mair, S.: Probabilistic movement models and zones of control. Mach. Learn. **108**(1), 127–147 (2019)
8. Bunker, R., Duy, V.N.L., Tabei, Y., Takeuchi, I., Fujii, K.: Multi-agent statistical discriminative sub-trajectory mining and an application to NBA basketball. J. Quant. Anal. Sports (2024)
9. Burke, B.: Deepqb: deep learning with player tracking to quantify quarterback decision-making & performance. In: Proceedings of the 2019 MIT Sloan Sports Analytics Conference (2019)
10. Cao, A., Xie, X., Zhou, M., Zhang, H., Mingliang, X., Yingcai, W.: Action-evaluator: a visualization approach for player action evaluation in soccer. IEEE Trans. Visual Comput. Graph. **30**(1), 880–890 (2024)
11. Capellera, G., Ferraz, L., Rubio, A., Agudo, A., Moreno-Noguer, F.: Footbots: a transformer-based architecture for motion prediction in soccer (2024). arXiv:2406.19852
12. Cervone, D., D'Amour, A., Bornn, L., Goldsberry, K.: Pointwise: predicting points and valuing decisions in real time with NBA optical tracking data. In: Proceedings of the MIT Sloan Sports Analytics Conference (2014)
13. Cervone, D., D'Amour, A., Bornn, L., Goldsberry, K.: A multiresolution stochastic process model for predicting basketball possession outcomes. J. Am. Stat. Assoc. **111**(514), 585–599 (2016)
14. Chang, Y.-H., Maheswaran, R., Kwok, S.J.J., Levy, T., Wexler, A., Squire, K.: Quantifying shot quality in the NBA. In: Proceedings of the MIT Sloan Sports Analytics Conference (2014)

15. Chen, C.-Y., Lai, W., Hsieh, H.-Y., Zheng, W.-H., Wang, Y.-S., Chuang, J.-H.: Generating defensive plays in basketball games. In: Proceedings of the 26th ACM International Conference on Multimedia, pp. 1580–1588 (2018)

16. Chen, C.-H., Liu, T.-L., Wang, Y.-S., Chu, H.-K., Tang, N.C., Mark Liao, H.-Y.: Spatio-temporal learning of basketball offensive strategies. In: Proceedings of the 23rd ACM International Conference on Multimedia, pp. 1123–1126 (2015)

17. Chen, X., Wang, W.-Y., Hu, Z., Chou, C., Hoang, L., Jin, K., Liu, M., Brantingham, P.J., Wang, W.: Professional basketball player behavior synthesis via planning with diffusion (2023). arXiv:2306.04090

18. Cohea, C, Payton, M.E.: Relationships between player actions and game outcomes in American football. Sportscience, **15**, 19–25 (2011)

19. Anthony Costa Constantinou and Norman Elliott Fenton: Solving the problem of inadequate scoring rules for assessing probabilistic football forecast models. J. Quant. Anal. Sports **8**(1), 1–14 (2012)

20. Anthony Costa Constantinou and Norman Elliott Fenton: Determining the level of ability of football teams by dynamic ratings based on the relative discrepancies in scores between adversaries. J. Quant. Anal. Sports **9**(1), 37–50 (2013)

21. Correia, V., Araujo, D., Craig, C., Passos, P.: Prospective information for pass decisional behavior in rugby union. Hum. Mov. Sci. **30**(5), 984–997 (2011)

22. Decroos, T., Bransen, L., Haaren, J.V., Davis, J.: Actions speak louder than goals: valuing player actions in soccer. In: Proceedings of the 25th ACM SIGKDD Conference on Knowledge Discovery and Data Mining, pp. 1851–1861 (2019)

23. Decroos, T, Davis, J.: Player vectors: Characterizing soccer players' playing style from match event streams. In: Joint European Conference on Machine Learning and Knowledge Discovery in Databases, pp. 569–584. Springer (2019)

24. Decroos, T., Haaren, J.V., Davis, J.: Automatic discovery of tactics in spatio-temporal soccer match data. In: ACM SIGKDD International Conference on Knowledge Discovery & Data Mining, pp. 223–232 (2018)

25. Deshpande, S.K., Evans, K.: Expected hypothetical completion probability. J. Quant. Anal. Sports **16**(2), 85–94 (2020)

26. Dick, U., Link, D., Brefeld, U.: Who can receive the pass? A computational model for quantifying availability in soccer. Data Min. Knowl. Disc. **36**(3), 987–1014 (2022)

27. Dick, U., Tavakol, M., Brefeld, U.: Rating player actions in soccer. Front. Sports Active Living **3**, 174 (2021)

28. Ding, N., Takeda, K., Jin, W., Bei, Y., Fujii, K.: Estimation of control area in badminton doubles with pose information from top and back view drone videos. Multimed. Tools Appl. **83**(8), 24777–24793 (2024)

29. Dixon, M.J., Coles, S.G.: Modelling association football scores and inefficiencies in the football betting market. J. R. Stat. Soc. Ser. C (Appl. Stat.) **46**(2), 265–280 (1997)

30. Dubitzky, W., Lopes, P., Davis, J., Berrar, D.: The open international soccer database for machine learning. Mach. Learn. **108**, 9–28 (2019)

31. Fassmeyer, D., Anzer, G., Bauer, P., Brefeld, U.: Toward automatically labeling situations in soccer. Front. Sports Active Living **3**, 725431 (2021)

32. Fassmeyer, D., Fassmeyer, P., Brefeld, U.: Semi-supervised generative models for multiagent trajectories. Adv. Neural. Inf. Process. Syst. **35**, 37267–37281 (2022)

33. Felsen, P., Lucey, P., Ganguly, S.: Where will they go? Predicting fine-grained adversarial multi-agent motion using conditional variational autoencoders. In: Proceedings of the European Conference on Computer Vision (ECCV), pp. 732–747 (2018)

34. Fernández, J, Bornn, L.: Wide open spaces: a statistical technique for measuring space creation in professional soccer. In: 12th MIT Sloan Sports Analytics Conference (2018)

35. Fernández, J., Bornn, L., Cervone, D.: Decomposing the immeasurable sport: a deep learning expected possession value framework for soccer. In: Proceedings of the MIT Sloan Sports Analytics Conference (2019)

36. Forcher, L., Altmann, S., Forcher, L., Jekauc, D., Kempe, M.: The use of player tracking data to analyze defensive play in professional soccer-a scoping review. Int. J. Sports Sci. Coach. **17**(6), 1567–1592 (2022)

37. Fujii, K., Inaba, Y., Kawahara, Y.: Koopman spectral kernels for comparing complex dynamics: application to multiagent sport plays. In: European Conference on Machine Learning and Knowledge Discovery in Databases (ECML-PKDD'17), pp. 127–139. Springer (2017)

38. Fujii, K., Isaka, T., Kouzaki, M., Yamamoto, Y.: Mutual and asynchronous anticipation and action in sports as globally competitive and locally coordinative dynamics. Sci. Rep. **5** (2015)

39. Fujii, K., Kawasaki, T., Inaba, Y., Kawahara, Y.: Prediction and classification in equation-free collective motion dynamics. PLoS Comput. Biol. **14**(11), e1006545 (2018)

40. Fujii, K., Takeishi, N., Hojo, M., Inaba, Y., Kawahara, Y.: Physically-interpretable classification of network dynamics for complex collective motions. Sci. Rep. **10**(3005) (2020)

41. Fujii, K., Takeishi, N., Kawahara, Y., Takeda, K.: Decentralized policy learning with partial observation and mechanical constraints for multiperson modeling. Neural Netw. **171**, 40–52 (2024)

42. Fujii, K., Takeuchi, K., Kuribayashi, A., Takeishi, N., Kawahara, Y., Takeda, K.: Estimating counterfactual treatment outcomes over time in complex multi-agent scenarios. IEEE Trans. Neural Netw. Learn. Syst. 1–15 (2024)

43. Fujii, K., Yamashita, D., Yoshioka, S., Isaka, T., Kouzaki, M.: Strategies for defending a dribbler: categorisation of three defensive patterns in 1-on-1 basketball. Sports Biomech, **13**(3), 204–214 (2014)

44. Fujii, K., Yokoyama, K., Koyama, T., Rikukawa, A., Yamada, H., Yamamoto, Y.: Resilient help to switch and overlap hierarchical subsystems in a small human group. Sci. Rep. **6** (2016)

45. Fujii, K., Yoshihara, Y., Matsumoto, Y., Tose, K., Takeuchi, H., Isobe, M., Mizuta, H., Maniwa, D., Okamura, T., Murai, T., et al.: Cognition and interpersonal coordination of patients with schizophrenia who have sports habits. PLoS ONE **15**(11), e0241863 (2020)

46. Fujimura, A., Sugihara, K.: Geometric analysis and quantitative evaluation of sport teamwork. Syst. Comput. Jpn. **36**(6), 49–58 (2005)

47. Gibbs, C.P., Elmore, R., Fosdick, B.K.: The causal effect of a timeout at stopping an opposing run in the nba. The Ann. Appl. Stat. **16**(3), 1359–1379 (2022)

48. Goes, F.R., Kempe, M., Meerhoff, L.A., Lemmink, K.A.P.M.: Not every pass can be an assist: a data-driven model to measure pass effectiveness in professional soccer matches. Big Data, **7**(1), 57–70 (2019)

49. Goldsberry, K.: Courtvision: new visual and spatial analytics for the nba. In: 2012 MIT Sloan Sports Analytics Conference, vol. 9, pp. 12–15 (2012)

50. Gómez, M.Á., Mitrotasios, M., Armatas, V., Lago-Peñas, C.: Analysis of playing styles according to team quality and match location in greek professional soccer. Int. J. Perform. Anal. Sport **18**(6), 986–997 (2018)

51. Graber, C., Schwing, A.G.: Dynamic neural relational inference. In: Proceedings of the IEEE/CVF Conference on Computer Vision and Pattern Recognition (CVPR), pp. 8513–8522 (2020)

52. Groll, A., Ley, C., Schauberger, G., Van Eetvelde, H.: A hybrid random forest to predict soccer matches in international tournaments. J. Quant. Anal. Sports **15**(4), 271–287 (2019)

53. Guangyu, L., Bo, J., Hao, Z., Zhengping, C., Yan, L.: Generative attention networks for multi-agent behavioral modeling. In: Thirty-Fourth AAAI Conference on Artificial Intelligence (2020)

54. Hauri, S., Vucetic, S.: Group activity recognition in basketball tracking data–neural embeddings in team sports (nets). In: ECAI 2023, pp. 1012–1019. IOS Press (2023)

55. Helbing, D., Molnar, P.: Social force model for pedestrian dynamics. Phys. Rev. E **51**(5), 4282 (1995)

56. Hobbs, J., Power, P., Sha, L.: Quantifying the value of transitions in soccer via spatiotemporal trajectory clustering. In: Proceedings of the MIT Sloan Sports Analytics Conference (2018)

57. Hojo, M., Fujii, K., Inaba, Y., Motoyasu, Y., Kawahara, Y.: Automatically recognizing strategic cooperative behaviors in various situations of a team sport. PLoS ONE **13**(12), e0209247 (2018)

58. Hojo, M., Fujii, K., Kawahara, Y.: Analysis of factors predicting who obtains a ball in basketball rebounding situations. Int. J. Perform. Anal. Sport, pp. 1–14 (2019)
59. Hoshen, Y.: Vain: attentional multi-agent predictive modeling. In: Advances in Neural Information Processing Systems, vol. 30, pp. 2701–2711 (2017)
60. Hsieh, H.-Y., Chen, C.-Y., Wang, Y.-S., Chuang, J.-H.: BasketballGAN: generating basketball play simulation through sketching. In: Proceedings of the 27th ACM International Conference on Multimedia, pp. 720–728 (2019)
61. Hughes, H., Horton, M., Wei, F., Stokl, M., Gammulle, H., Fookes, C., Sridharan, S., Pedagadi, S., Lucey, P.: Approaching in-venue quality tracking from broadcast video using generative ai. In: Proceedings of the MIT Sloan Sports Analytics Conference (2024)
62. Ishida, A., Takayanagi, M., Hoshina, I., Iwayama, K.: Bayesian credible possession based player performance evaluation in basketball. Keisankitoukeigaku (J. Jpn. Soc. Comput. Stat.) **36**(2), 99–126 (2023)
63. Ivanovic, B., Schmerling, E., Leung, K., Pavone, M.: Generative modeling of multimodal multi-human behavior. In: 2018 IEEE/RSJ International Conference on Intelligent Robots and Systems, pp. 3088–3095. IEEE (2018)
64. Iwashita, S., Scott, A., Umemoto, R., Ding, N., Fujii, K.: Space evaluation based on pitch control using drone video in ultimate (2024). arXiv preprint
65. Kalman, S., Bosch, J.: NBA lineup analysis on clustered player tendencies: a new approach to the positions of basketball & modeling lineup efficiency of soft lineup aggregates. In: 14th Annual MIT Sloan Sports Analytics Conference (2020)
66. Kanda, S., Takeuchi, K., Fujii, K., Tabei, Y.: Succinct trit-array trie for scalable trajectory similarity search. In: Proceedings of the 28th International Conference on Advances in Geographic Information Systems, pp. 518–529 (2020)
67. Ke, G., Meng, Q., Finley, T., Wang, T., Chen, W., Ma, W., Ye, Q., Liu, T.-Y.: Lightgbm: a highly efficient gradient boosting decision tree. Adv. Neural. Inf. Process. Syst. **30**, 3146–3154 (2017)
68. Kempe, M., Grunz, A., Memmert, D.: Detecting tactical patterns in basketball: comparison of merge self-organising maps and dynamic controlled neural networks. Eur. J. Sport Sci. **15**(4), 249–255 (2015)
69. Kijima, A., Yokoyama, K., Shima, H., Yamamoto, Y.: Emergence of self-similarity in football dynamics. Eur. Phys. J. B **87**(2), 41 (2014)
70. Kim, J., James, N., Parmar, N., Ali, B., Vučković, G.: The attacking process in football: a taxonomy for classifying how teams create goal scoring opportunities using a case study of crystal palace fc. Front. Psychol. **10**, 2202 (2019)
71. Kipf, T., Fetaya, E., Wang, K.-C., Welling, M., Zemel, R.: Neural relational inference for interacting systems. In: International Conference on Machine Learning, pp. 2688–2697 (2018)
72. Knoll, J., Stübinger, J.: Machine-learning-based statistical arbitrage football betting. KI-Künstliche Intelligenz **34**(1), 69–80 (2020)
73. Kobayashi, R., Umemoto, R., Takeda, K., Fujii, K.: Score prediction using multiple object tracking for analyzing movements in 2-vs-2 handball. In: 2023 IEEE 12th Global Conference on Consumer Electronics (GCCE), pp. 946–947. IEEE (2023)
74. Kono, R., Fujii, K.: Mathematical models for off-ball scoring prediction in basketball. In: International Workshop on Machine Learning and Data Mining for Sports Analytics. Springer (2024)
75. Lago-Peñas, C., Gómez-Ruano, M., Yang, G.: Styles of play in professional soccer: an approach of the Chinese soccer super league. Int. J. Perform. Anal. Sport **17**(6), 1073–1084 (2017)
76. Lamas, L., Santana, F., Heiner, M., Ugrinowitsch, C., Fellingham, G.: Modeling the offensive-defensive interaction and resulting outcomes in basketball. PLoS One **10**(12), 1–14, 12 (2015)
77. Le, H.M., Yue, Y., Carr, P., Lucey, P.: Coordinated multi-agent imitation learning. In: Proceedings of the 34th International Conference on Machine Learning, Vol. 70, pp. 1995–2003. JMLR (2017). (JMLR. org, 2017)

78. Link, D., Lang, S., Seidenschwarz, P.: Real time quantification of dangerousity in football using spatiotemporal tracking data. PLoS ONE **11**(12), e0168768 (2016)
79. Llana, S., Madrero, P., Fernández, J., Barcelona, F.C.: The right place at the right time: advanced off-ball metrics for exploiting an opponent's spatial weaknesses in soccer. In: Proceedings of the MIT Sloan Sports Analytics Conference (2020)
80. Lucey, P., Bialkowski, A., Monfort, M., Carr, P., Matthews, I.: Quality vs quantity: improved shot prediction in soccer using strategic features from spatiotemporal data. In: Proceedings of the MIT Sloan Sports Analytics Conference (2014)
81. Lucey, P., Oliver, D., Carr, P., Roth, J., Matthews, I.: Assessing team strategy using spatiotemporal data. In: Proceedings of the 19th ACM SIGKDD International Conference on Knowledge Discovery and Data Mining, pp. 1366–1374 (2013)
82. Ma, J., Guo, R., Chen, C., Zhang, A., Li, J.: Deconfounding with networked observational data in a dynamic environment. In: Proceedings of the 14th ACM International Conference on Web Search and Data Mining, pp. 166–174 (2021)
83. Maher, M.J.: Modelling association football scores. Statistica Neerlandica **36**(3), 109–118 (1982)
84. Martens, F., Dick, U., Brefeld, U.: Space and control in soccer. Front. Sports Active Living **3**, 676179 (2021)
85. Marty, R.: High-resolution shot capture reveals systematic biases and an improved method for shooter evalutation. In: Proceedings of the 2018 MIT Sloan Sports Analytics Conference (2018)
86. McHale, I., Scarf, P.: Modelling soccer matches using bivariate discrete distributions with general dependence structure. Stat. Neerl. **61**(4), 432–445 (2007)
87. McHale, I.G., Scarf, P.A., Folker, D.E.: On the development of a soccer player performance rating system for the English premier league. Interfaces **42**(4), 339–351 (2012)
88. McIntyre, A., Brooks, J., Guttag, J., Wiens, J.: Recognizing and analyzing ball screen defense in the NBA. In: Proceedings of the MIT Sloan Sports Analytics Conference, pp. 11–12 (2016)
89. McQueen, A., Wiens, J., Guttag, J.: Automatically recognizing on-ball screens. In: Proceedings of the MIT Sloan Sports Analytics Conference (2014)
90. Mendes-Neves, T., Meireles, L., Mendes-Moreira, J.: Estimating player performance in different contexts using fine-tuned large events models (2024). arXiv:2402.06815
91. Mendes-Neves, T., Meireles, L., Mendes-Moreira, J.: Forecasting events in soccer matches through language (2024). arXiv:2402.06820
92. Merckx, S., Robberechts, P., Euvrard, Y., Davis, J.: Measuring the effectiveness of pressing in soccer. In: Workshop on Machine Learning and Data Mining for Sports Analytics (2021)
93. Merlin, M., Cunha, S.A., Moura, F.A., da Silva Torres, R., Gonçalves, B., Sampaio, J.: Exploring the determinants of success in different clusters of ball possession sequences in soccer. Res. Sports Med. **28**(3), 339–350 (2020)
94. Miller, A., Bornn, L., Adams, R., Goldsberry, K.: Factorized point process intensities: a spatial analysis of professional basketball. In: International Conference on Machine Learning, pp. 235–243 (2014)
95. Miller, A.C., Bornn, L.: Possession sketches: mapping NBA strategies. In: Proceedings of MIT Sloan Sports Analytics Conference (2017)
96. Mortensen, J, Bornn, L.: From Markov models to Poisson point processes: modeling movement in the NBA. In: Proceedings of the 2019 MIT Sloan Sports Analytics Conference (2019)
97. Müller, O., Caron, M., Döring, M., Heuwinkel, T., Baumeister, J.: Pivot: a parsimonious end-to-end learning framework for valuing player actions in handball using tracking data. In: International Workshop on Machine Learning and Data Mining for Sports Analytics, pp. 116–128. Springer (2021)
98. Mures, O.A., Taibo, J., Padrón, E.J., Iglesias-Guitian, J.A.: Playnet: real-time handball play classification with Kalman embeddings and neural networks. Vis. Comput. **40**(4), 2695–2711 (2024)

99. Nakahara, H., Takeda, K., Fujii, K.: Estimating the effect of team hitting strategies using counterfactual virtual simulation in baseball. Int. J. Comput. Sci. Sport **22**(1), 1–12 (2023)
100. Nakahara, H., Takeda, K., Fujii, K.: Pitching strategy evaluation via stratified analysis using propensity score. J. Quant. Anal. Sports **19**(2), 91–102 (2023)
101. Nakahara, H., Tsutsui, K., Takeda, K., Fujii, K.: Action valuation of on-and off-ball soccer players based on multi-agent deep reinforcement learning. IEEE Access **11**, 131237–131244 (2023)
102. Narizuka, T., Yamazaki, Y., Takizawa, K.: Space evaluation in football games via field weighting based on tracking data. Sci. Rep. **11**(1), 5509 (2021)
103. Nistala, A.: Using deep learning to understand patterns of player movement in basketball. Ph.D thesis, Massachusetts Institute of Technology (2018)
104. Niu, Z., Gao, X., Tian, Q.: Tactic analysis based on real-world ball trajectory in soccer video. Pattern Recogn. **45**(5), 1937–1947 (2012)
105. Omidshafiei, S., Hennes, D., Garnelo, M., Wang, Z., Recasens, A., Tarassov, E., Yang, Y., Elie, R., Connor, J.T., Muller, P., et al.: Multiagent off-screen behavior prediction in football. Sci. Rep. **12**(1), 8638 (2022)
106. Papalexakis, E., Pelechrinis, K.: thoops: a multi-aspect analytical framework for spatio-temporal basketball data. In: Proceedings of the 27th ACM International Conference on Information and Knowledge Management, pp. 2223–2232 (2018)
107. Pappalardo, L., Cintia, P., Ferragina, P., Massucco, E., Pedreschi, D., Giannotti, F.: Playerank: data-driven performance evaluation and player ranking in soccer via a machine learning approach. ACM Trans. Intell. Syst. Technol. (TIST) **10**(5), 1–27 (2019)
108. Pappalardo, L., Cintia, P., Rossi, A., Massucco, E., Ferragina, P., Pedreschi, D., Giannotti, F.: A public data set of spatio-temporal match events in soccer competitions. Sci. Data **6**(1), 1–15 (2019)
109. Park, J.Y., Carr, K., Zheng, S., Yue, Y., Yu, R.: Multiresolution tensor learning for efficient and interpretable spatial analysis. In: International Conference on Machine Learning, pp. 7499–7509. PMLR (2020)
110. Perše, M., Kristan, M., Kovačič, S., Vučković, G., Perš, J.: A trajectory-based analysis of coordinated team activity in a basketball game. Comput. Vis. Image Underst. **113**(5), 612–621 (2009)
111. Power, P., Hobbs, J., Ruiz, H., Wei, X., Lucey, P.: Mythbusting set-pieces in soccer. In: Proceedings of the MIT Sloan Sports Analytics Conference (2018)
112. Power, P., Ruiz, H., Wei, X., Lucey, P.: Not all passes are created equal: Objectively measuring the risk and reward of passes in soccer from tracking data. In: Proceedings of the 23rd ACM SIGKDD International Conference on Knowledge Discovery and Data Mining, pp. 1605–1613 (2017)
113. Rahimian, P., da Silva Guerra Gomes, D.G., Berkovics, F., Toka, L.: Let's penetrate the defense: a machine learning model for prediction and valuation of penetrative passes. In: International Workshop on Machine Learning and Data Mining for Sports Analytics, pp. 41–52. Springer (2022)
114. Rahimian, P., Kim, H., Schmid, M., Toka, L.: Pass receiver and outcome prediction in soccer using temporal graph networks. In: International Workshop on Machine Learning and Data Mining for Sports Analytics, pp. 52–63. Springer (2023)
115. Razali, M.N., Mustapha, A., Mostafa, S.A., Gunasekaran, S.S.: Football matches outcomes prediction based on gradient boosting algorithms and football rating system. Hum. Factors Softw. Syst. Eng. **61**, 57 (2022)
116. Razali, N., Mustapha, A., Arbaiy, N., Lin, P.-C.: Deep learning for football outcomes prediction based on football rating system. In: AIP Conference Proceedings, vol. 2644. AIP Publishing (2022)
117. Robberechts, P.: Valuing the art of pressing. In: StatsBomb Innovation in Football Conference (2019)
118. Robberechts, P., Van Roy, M., Davis, J.: un-xpass: measuring soccer player's creativity. In: Proceedings of the 29th ACM SIGKDD Conference on Knowledge Discovery and Data Mining, pp. 4768–4777 (2023)

119. Rudolph, Y., Brefeld, U.: Modeling conditional dependencies in multiagent trajectories. In: Camps-Valls, G., Ruiz, F.J.R., Valera, I. (eds.) Proceedings of The 25th International Conference on Artificial Intelligence and Statistics. Proceedings of Machine Learning Research, , vol. 151, pp. 10518–10533. PMLR (28–30 Mar 2022)

120. Rudolph, Y., Brefeld, U.: Masked autoencoder pretraining for event classification in elite soccer. In: International Workshop on Machine Learning and Data Mining for Sports Analytics, pp. 24–35. Springer (2023)

121. Sampaio, J., McGarry, T., Calleja-González, J., Jiménez Sáiz, S., Schelling i del Alcázar, X., Balciunas, M.: Exploring game performance in the national basketball association using player tracking data. PLoS One **10**(7), e0132894 (2015)

122. Schmid, M., Blauberger, P., Lames, M.: Simulating defensive trajectories in American football for predicting league average defensive movements. Front. Sports Active Living **3**, 669845 (2021)

123. Seidl, T., Cherukumudi, A., Hartnett, A., Carr, P., Lucey, P.: Bhostgusters: realtime interactive play sketching with synthesized nba defenses. In: Proceedings of the MIT Sloan Sports Analytics Conference (2018)

124. Sha, L.: Representing and predicting multi-agent data in adversarial team sports. Ph.D thesis, Queensland University of Technology (2018)

125. Sha, L., Lucey, P., Yue, Y., Carr, P., Rohlf, C., Matthews, I.: Chalkboarding: a new spatiotemporal query paradigm for sports play retrieval. In: International Conference on Intelligent User Interfaces, pp. 336–347 (2016)

126. Simpson, I., Beal, R.J., Locke, D., Norman, T.J.: Seq2event: learning the language of soccer using transformer-based match event prediction. In: Proceedings of the 28th ACM SIGKDD Conference on Knowledge Discovery and Data Mining, pp. 3898–3908 (2022)

127. Spearman, W.: Beyond expected goals. In: Proceedings of the 12th MIT Sloan Sports Analytics Conference, pp. 1–17 (2018)

128. Stavropoulos, N., Papadopoulou, A., Kolias, P.: Evaluating the efficiency of off-ball screens in elite basketball teams via second-order Markov modelling. Mathematics **9**(16), 1991 (2021)

129. Sun, C., Karlsson, P., Wu, J., Tenenbaum, J.B., Murphy, K.: Predicting the present and future states of multi-agent systems from partially-observed visual data. In: International Conference on Learning Representations (2019)

130. Supola, B., Hoch, T., Baca, A.: Modeling the offensive-defensive interaction and resulting outcomes in basketball. PLoS ONE **18**(2), e0281467 (2023)

131. Takeuchi, K., Imaizumi, M., Kanda, S., Tabei, Y., Fujii, K., Yoda, K., Ishihata, M., Maekawa, T.: Fréchet kernel for trajectory data analysis. In: Proceedings of the 29th International Conference on Advances in Geographic Information Systems, pp. 221–224 (2021)

132. Taki, T., Hasegawa, J., Fukumura, T.: Development of motion analysis system for quantitative evaluation of teamwork in soccer games. In: Proceedings of 3rd IEEE International Conference on Image Processing, vol. 3, pp. 815–818. IEEE (1996)

133. Teranishi, M., Fujii, K., Takeda, K.: Trajectory prediction with imitation learning reflecting defensive evaluation in team sports. In: 2020 IEEE 9th Global Conference on Consumer Electronics (GCCE), pp. 124–125. IEEE (2020)

134. Teranishi, M., Tsutsui, K., Takeda, K., Fujii, K.: Evaluation of creating scoring opportunities for teammates in soccer via trajectory prediction. In: International Workshop on Machine Learning and Data Mining for Sports Analytics, pp. 53–73. Springer (2022)

135. Toda, K., Teranishi, M., Kushiro, K., Fujii, K.: Evaluation of soccer team defense based on prediction models of ball recovery and being attacked: a pilot study. PLoS ONE **17**(1), e0263051 (2022)

136. Toumi, A., Lopez, M.: From grapes and prunes to apples and apples: using matched methods to estimate optimal zone entry decision-making in the national hockey league. In: Carnegie Mellon Sports Analytics Conference 2019 (2019)

137. Travassos, B., Araújo, D., Duarte, R., McGarry, T.: Spatiotemporal coordination behaviors in futsal (indoor football) are guided by informational game constraints. Hum. Mov. Sci. **31**(4), 932–945 (2012)

138. Umemoto, R., Fujii, K.: Evaluation of team defense positioning by computing counterfactuals using statsbomb 360 data. In: StatsBomb Conference (2023)
139. van der Maaten, L., Hinton, G.: Visualizing data using t-SNE. J. Mach. Learn. Res. **9**, 2579–2605 (2008)
140. Van Roy, M., Cascioli, L., Davis, J.: Etsy: a rule-based approach to event and tracking data synchronization. In: International Workshop on Machine Learning and Data Mining for Sports Analytics, pp. 11–23. Springer (2023)
141. Van Roy, M., Robberechts, P., Davis, J.: Optimally disrupting opponent build-ups. In: Proceedings of the 2021 StatsBomb Conference, London, UK, pp. 1–16 (2021)
142. Van Roy, M., Robberechts, P., Decroos, T., Davis, J.: Valuing on-the-ball actions in soccer: a critical comparison of xt and vaep. In: Proceedings of the AAAI-20 Workshop on Artifical Intelligence in Team Sports. AI in Team Sports Organising Committee (2020)
143. Vilar, L., Araújo, D., Davids, K., Correia, V., Esteves, P.T.: Spatial-temporal constraints on decision-making during shooting performance in the team sport of futsal. J. Sports Sci. **31**(8), 840–846 (2013)
144. David Michael Vock and Laura Frances Boehm Vock: Estimating the effect of plate discipline using a causal inference framework: an application of the g-computation algorithm. J. Quant. Anal. Sports **14**(2), 37–56 (2018)
145. Wang, K.-C., Zemel, R.: Classifying nba offensive plays using neural networks. In: Proceedings of MIT Sloan Sports Analytics Conference (2016)
146. Wang, Q., Zhu, H., Hu, W., Shen, Z., Yao, Y.: Discerning tactical patterns for professional soccer teams: an enhanced topic model with applications. In: Proceedings of the 21th ACM SIGKDD International Conference on Knowledge Discovery and Data Mining, pp. 2197–2206 (2015)
147. Wang, X., Tang, Z., Jianchong Shao, Z., Robertson, S., Gómez, M.-Á., Zhang, S.: Hooptransformer: advancing NBA offensive play recognition with self-supervised learning from player trajectories. Sports Med. 1–11 (2024)
148. Wang, Z., Veličković, P., Hennes, D., Tomašev, N., Prince, L., Kaisers, M., Bachrach, Y., Elie, R., Wenliang, L.K., Piccinini, F., et al.: Tacticai: an ai assistant for football tactics. Nat. Commun. **15**(1), 1906 (2024)
149. Wang, Z., Long, C., Cong, G., Ju, C.: Effective and efficient sports play retrieval with deep representation learning. In: Proceedings of the 25th ACM SIGKDD International Conference on Knowledge Discovery & Data Mining, pp. 499–509 (2019)
150. Wei, Z., Zhu, X., Dai, B., Lin, D.: Rethinking trajectory prediction via team game (2022). arXiv:2210.08793
151. Williams, A.M.: Perceptual skill in soccer: implications for talent identification and development. J. Sports Sci. **18**(9), 737–750 (2000)
152. Wu, L.Y., Danielson, A.J., Joan Hu, X., Swartz, T.B.: A contextual analysis of crossing the ball in soccer. J. Quant. Anal. Sports **17**(1), 57–66 (2021)
153. Yihong, W., Deng, D., Xie, X., He, M., Jie, X., Zhang, H., Zhang, H., Yingcai, W.: Obtracker: visual analytics of off-ball movements in basketball. IEEE Trans. Visual Comput. Graph. **29**(1), 929–939 (2023)
154. Yam, D.R., Lopez, M.J.: What was lost? A causal estimate of fourth down behavior in the national football league. J. Sports Anal. **5**(3), 153–167 (2019)
155. Yamada, K., Fujii, K.: Offensive lineup analysis in basketball with clustering players based on shooting style and offensive role (2024). arXiv:2403.13821
156. Yamamoto, Y., Yokoyama, K.: Common and unique network dynamics in football games. PLoS ONE **6**(12), e29638 (2011)
157. Yeh, R.A., Schwing, A.G., Huang, J., Murphy, K.: Diverse generation for multi-agent sports games. In: The IEEE Conference on Computer Vision and Pattern Recognition (CVPR), pp. 4610–4619 (2019)
158. Yeung, C., Bunker, R., Umemoto, R., Fujii, K.: Evaluating soccer match prediction models: a deep learning approach and feature optimization for gradient-boosted trees. Mach. Learn. 1–24 (2024)

159. Yeung, C., Fujii, K.: A strategic framework for optimal decisions in football 1-vs-1 shot-taking situations: an integrated approach of machine learning, theory-based modeling, and game theory. Complex Intell. Syst. 1–20 (2024)
160. Yeung, C., Ide, K., Fujii, K.: Autosoccerpose: automated 3d posture analysis of soccer shot movements. In: Proceedings of the IEEE/CVF Conference on Computer Vision and Pattern Recognition, pp. 3214–3224 (2024)
161. Yeung, C.C.K., Sit, T., Fujii, K.: Transformer-based neural marked spatio temporal point process model for football match events analysis (2023). arXiv:2302.09276
162. Yokoyama, K., Yamamoto, Y.: Three people can synchronize as coupled oscillators during sports activities. PLoS Comput. Biol. **7**(10), e1002181 (2011)
163. Yue, Y., Lucey, P., Carr, P., Bialkowski, A., Matthews, I.: Learning fine-grained spatial models for dynamic sports play prediction. In: 2014 IEEE International Conference on Data Mining, pp. 670–679. IEEE (2014)
164. Yurko, R., Matano, F., Richardson, L.F., Granered, N., Pospisil, T., Pelechrinis, K., Ventura, S.L.: Going deep: models for continuous-time within-play valuation of game outcomes in American football with tracking data. J. Quant. Anal. Sports **16**(2), 163–182 (2020)
165. Zhan, E., Zheng, S., Yue, Y., Sha, L., Lucey, P.: Generating multi-agent trajectories using programmatic weak supervision. In: International Conference on Learning Representations (2019)
166. Zhang, Z., Bunker, R., Takeda, K., Fujii, K.: Multi-agent deep-learning based comparative analysis of team sport trajectories. IEEE Access **11**, 43305–43315 (2023)
167. Zheng, S., Yue, Y., Hobbs, J.: Generating long-term trajectories using deep hierarchical networks. In: Advances in Neural Information Processing Systems, vol. 29, pp. 1543–1551 (2016)
168. Ziyi, Z., Takeda, K., Fujii, K.: Cooperative play classification in team sports via semi-supervised learning. Int. J. Comput. Sci. Sport **21**(1), 111–121 (2022)

Chapter 4
Potential Play Evaluation with Learning-Based Agent Modeling

Abstract This chapter explores the potential of learning-based agent models in play evaluation for sports analytics. It begins by discussing the fundamentals of agent modeling, including key concepts in this field, such as the Markov decision process, reinforcement learning, and game theory, and including several simulation platforms. Next, the inverse approach for player and team evaluation from data and the forward approach for virtual simulations are introduced, and the advantages of using learning-based methods over traditional simulation techniques are highlighted. These approaches collectively contribute to the advancement of modeling complex scenarios, evaluating different tactical choices, and suggesting optimal actions in sports. Finally, this chapter examines the technical and practical challenges associated with these methods, as well as future research opportunities in the field. This chapter may provide us with a comprehensive understanding of how learning-based agent modeling can enhance play evaluation and contribute to the advancement of sports analytics.

Keywords Agent modeling · Reinforcement learning · Game theory · Simulation · Multi-agent systems

4.1 Introduction

The previous Chap. 3 explores predictive analysis and play evaluation using machine learning, focusing primarily on supervised/unsupervised learning or pattern-based solutions. These approaches leverage historical data to identify patterns and predict outcomes, providing an intelligent framework for understanding player behavior and their interactions. However, they often lack controllability when considering new scenarios and may not capture the complex decision making of players. In contrast, learning-based agent modeling offers a dynamic approach that can simulate and evaluate their decision making and their interactions, providing deeper insights into players' potential plays and their evaluations.

The application of agent modeling in sports analytics presents several challenges. One of the primary difficulties is accurately representing the diverse behaviors

© The Author(s) 2025
K. Fujii, *Machine Learning in Sports*,
SpringerBriefs in Computer Science, https://doi.org/10.1007/978-981-96-1445-5_4

and interactions of players as multi-agent systems. Traditional rule-based methods require manually defining the rules and parameters for player movements, which can be time-consuming and inflexible. These approaches struggle to adapt to different sports or new scenarios without extensive manual adjustments. A data-driven approach, utilizing machine learning, offers a potential solution by automatically learning from vast amounts of data. This allows for the creation of more adaptable and realistic models that can be generalized across various sports and scenarios, enhancing the efficiency and effectiveness of sports simulators. If a high-quality sports simulator can be developed, it would be possible to simulate various real-world scenarios, allowing for the evaluation of different action choices and the suggestions of optimal actions.

This chapter explores the potential of learning-based agent models in play evaluation. The chapter begins by discussing the fundamentals of agent modeling, including key concepts in this field, such as the Markov decision process (MDP), reinforcement learning (RL), game theory, and including several simulation platforms in Sect. 4.2. Next, Sects. 4.3 and 4.4 introduce the inverse approach for player and team evaluation from data and the forward approach for virtual simulations, respectively, and the advantages of using learning-based methods over traditional simulation techniques are highlighted. Finally, Sect. 4.5 examines the technical and practical challenges associated with these methods, as well as future research opportunities in the field.

4.2   Key Concepts of Agent Modeling

Agent modeling is a powerful approach in sports analytics, allowing for the simulation, evaluation, and suggestion of player behavior and interactions. This approach involves the creation of virtual agents that can mimic real-world players, enabling the verification of researchers' hypotheses through future prediction or in situations that cannot be directly measured. Agent modeling involves planning, which explicitly addresses the long-term movement goals of agents and computes policies or path hypotheses to achieve these goals. The key concepts of agent modeling are grounded in the principles of MDP and RL, which provide a mathematical framework for decision-making, described in Sect. 4.2.1. Next, various available platforms of sports simulators are introduced in Sect. 4.2.2, which facilitate the development and testing of these models. Then, traditional rule-based simulations and applications of game theory to team sports are introduced in Sects. 4.2.3 and 4.2.4, respectively. Finally, inverse and forward planning approaches are described for understanding the evolution and application of agent modeling in sports analytics.

4.2.1 *Markov Decision Process and Reinforcement Learning*

Here, we consider a sequential decision-making setting involving multiple agents interacting in a team sports environment. To facilitate our discussion, MDP is introduced as illustrated in Fig. 4.1, which provides a mathematical framework for decision-making in stochastic environments. Although real-world players do not necessarily follow an MDP, for simplicity, we use this model and assume that all aspects of the environment are fully observable.

A multi-agent MDP can be defined as a tuple $(K, S, A, \mathcal{T}, R, \gamma)$, where:

- K is the fixed number of agents;
- S is the set of states s;
- $A = [A_1, ..., A_K]$ represents the set of joint action $\mathbf{a} \in A$ (for a variable number of agents), and A_k is the set of joint actions, and A_k is the set of local actions a_k that agent k can take;
- $\mathcal{T}(s'|s, \mathbf{a}) : S \times A \times S \rightarrow [0, 1]$ is the transition model for all agents;
- $R = [R_1, ..., R_K] : S \times A \rightarrow \mathbb{R}^K$ is the joint reward function;
- $\gamma \in (0, 1]$ is the discount factor.

In the MDP, the goal of each agent is to maximize the expected total reward over time by selecting actions based on the current state of the environment. The transition model $\mathcal{T}$ determines the probability of moving from one state to another given the agents' actions, and the reward function R assigns a numerical reward to each state-action pair, guiding the agents' decision-making process.

RL is a type of machine learning where agents learn to make decisions by interacting with their environment and receiving rewards or penalties [55]. The aim is to learn a policy that maximizes cumulative reward. Recent RL algorithms, including deep RL, have advanced the field significantly by enabling agents to learn from high-dimensional sensory inputs and complex environments (e.g., [36]), and by integrating various strategies to balance exploration and exploitation and adapt to dynamic environments (e.g., [3]).

In the context of multi-agent reinforcement learning (MARL), several additional challenges arise compared to single-agent RL (e.g., [72]). One key challenge is the non-stationary environment caused by the presence of multiple learning agents. Since each agent's actions and policies evolve over time, the environment from any single agent's perspective is constantly changing. This non-stationarity can make it difficult for agents to learn stable and effective policies.

Fig. 4.1 Conceptual
diagram of a MDP

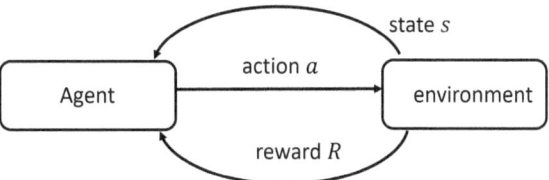

Another significant challenge in MARL is the coordination and cooperation among agents. In many multi-agent settings, agents must work together to achieve common goals or optimize joint rewards. This requires sophisticated mechanisms for competition and coordination, which are not typically required in single-agent RL. Designing algorithms that can facilitate effective cooperation and gain a competitive edge while ensuring each agent pursues its own objective is a complex task. Additionally, the presence of competition and conflicting objectives among agents necessitates anticipating and responding to opponents' strategies, a strategic interaction often modeled using game theory; balancing cooperative and competitive interactions, as well as developing strategies robust against adversarial behaviors, are critical challenges that will be further detailed in Sect. 4.2.4 and Chap. 5.

In the context of RL in team sports, due to the lack of real-world data and the difficulty of achieving tasks in a continuous state space, RL algorithms often start with simpler approaches, focusing first on defining the states, actions, transition models, and reward functions, which can be determined in either rule-based or learning-based manners. Some of the platforms discussed in the following Sect. 4.2.2 implement these elements using rule-based approaches but also support RL, allowing for the development and validation of learning-based agent models.

4.2.2 Simulation Platforms

Team sports simulation platforms have evolved significantly over the years, offering researchers and developers powerful tools to study and improve various aspects of sports movements and strategies. RoboCup [17], initiated as the Robot World Cup Initiative, represents a pioneering effort in bridging physical and virtual spaces for sports simulation. The initiative has expanded to include various leagues, including the 2D Soccer Simulation League (e.g., [1]), which is particularly noteworthy in the context of sports analytics. This league operates on a two-dimensional plane, abstracting complex physical interactions into a more manageable format. The 2D simulation environment shares a similar level of abstraction with event and tracking data in sports analytics, making it highly compatible with real-world sports analysis techniques. This synergy allows researchers to apply insights gained from 2D simulations to actual sports scenarios and vice versa.

In recent years, several simulation platforms have emerged for RL. Google research football (GFootball) [18] provides a comprehensive environment for RL in soccer, offering various difficulty levels and scenarios. Based on this, we created an original soccer environment called NFootball [15] because in GFootball [18], the transition algorithms are difficult to customize, and some commands (e.g., pass) did not work well within the intended timings. NFootball has a simple soccer environment and all algorithms are written in Python, and then transparent based on MAPE environments [32, 60]. The Simple Team Sports Simulator [52], developed by Electronic Arts, offers a flexible framework for simulating generic team sports, allowing researchers to model and analyze various game dynamics. A data-driven simulator for assessing decision-making using soccer event data has also been developed [35].

While publicly available team sports simulators offer valuable research opportunities, it is worth noting that some popular commercial games can be used but directly unavailable for academic purposes. Many esports titles like EA Sports FIFA, NBA 2K, and eFootball, have sophisticated simulation engines that could potentially benefit research, but their proprietary nature limits their use outside of commercial contexts.[1] Additionally, some researchers used a 3-vs-3 basketball simulator called Fever Basketball Defense [57], and a different platform was developed [34], both of which are currently unavailable.

4.2.3   Traditional Rule-Based Simulations

In traditional rule-based methods, researchers manually establish the rules governing agent movements, such as approaching the ball or avoiding opponents. Model parameters, including player positions, speeds, and interactions, are either manually set or statistically estimated through regression models based on the knowledge of sports. For instance, the movements in a 3-vs-1 soccer possession task are often modeled using three virtual social forces: spatial, avoiding, and cooperative forces [70]. More complex rule-based approaches have been used to model pass probabilities [53] and the future trajectories of players [2] in actual soccer games. These rule-based methods offer a clear understanding of simulated behaviors because all rules are explicitly defined by the users. However, adapting these methods to different sports, such as transitioning from soccer to basketball, requires significant human effort and it is challenging to derive universal rules for multi-agent behaviors.

While comprehensive simulations of full-player games (e.g., 11-vs-11 soccer matches) are successfully conducted in projects like RoboCup or esports games, research at the academic level often focuses on simulating partial group movements, as seen in studies like the aforementioned examples. Numerous pedestrian simulators exist, but the complex rules of team sports make it burdensome for individual researchers to develop and publicly share detailed team sport simulators.

4.2.4   Game Theory

Game theory plays an essential role in the study of sports, offering a theoretical basis for understanding players' behavioral strategies. The game theory differs fundamentally from multi-agent MDP (MAMDP) and MARL by focusing on the strategic interactions among multiple players (e.g., discussed in [19]). While MAMDP and

---

[1] When the application is limited to depth estimation [22] or 3D mesh reconstruction [74], some game data has been made publicly available, contributing to the field of computer vision.

MARL primarily aim to maximize individual agent rewards through learning optimal policies within a defined environment, game theory emphasizes the equilibrium that arises from the interplay of strategies among rational players. In MAMDP, agents operate under a framework that includes state spaces, action spaces, and reward functions, and MARL extends this by allowing agents to learn and adapt their strategies over time through interaction with the environment. However, these approaches do not inherently account for the strategic dependencies between agents' actions. In contrast, game theory introduces the concept of Nash equilibrium, where each player's strategy is optimal given the strategies of all other players [39, 56]. This equilibrium condition ensures that no player can unilaterally improve their payoff by changing their strategy, thus providing a stable solution concept for analyzing strategic interactions in competitive and cooperative settings. The distinct focus of game theory on equilibrium and strategic interdependence sets it apart from the more individually focused optimization goals of MAMDP and MARL.

Comprehensive overviews of how game theory can be applied to team sports [23, 50] and specifically to soccer [61] have been provided. In team sports, for simplicity, game theory is often utilized for tactical choices in static settings with complete information. For example, the majority of existing game-theoretic studies in soccer and basketball have focused on simple scenarios such as penalty kicks in soccer [4, 5, 7, 8, 25, 40] and live shot [37], and shot selection in basketball [51], where the interaction is primarily between the shooter and one or no defender (goalkeeper), allowing for more precise analysis. For other team sports, volleyball [28], American football [10, 67], handball, and ice hockey [23] have been investigated.

In particular, Yeung and Fujii [69] employed Nash equilibrium to analyze the interaction strategies between the shooter and the closest defender in soccer shot-taking scenarios, using a Shooting Payoff Computation (SPC) framework as described in Sect. 3.5.1. The optimal strategy for the shooter was to pass the ball when the defender was in a blocking position, while the optimal strategy for the defender was to block the shot. This situation represents a Nash equilibrium, where neither player can improve their expected payoff by unilaterally changing their strategy.

Despite these advances, the potential benefits of game-theoretic analysis in team sports are limited unless the focus shifts from static set-piece analysis to more dynamic settings, which is discussed in the review of [61]. Addressing the complexities of scenarios involves overcoming significant challenges, such as the large number of active players, the exponential growth of the strategy space, and the variability in player trajectories. Additionally, the longer duration of plays in team sports compared to those in more controlled environments suggests that extensive-form analysis, which considers each player's knowledge, opportunities, and actions, may be more appropriate than the simpler normal-form approaches typically used for set pieces. More advanced topics are discussed in Sect. 5.1.

4.2.5 *Inverse and Forward Approaches in Agent Modeling*

Planning-based methods focus on addressing the long-term objectives of agents by computing policies or path hypotheses that allow agents to achieve these goals. Planning-based methods can be divided into two categories: inverse and forward approaches. In general, as illustrated in Fig. 1.5, a forward problem involves generating results (data) from known causes (models), whereas an inverse problem involves estimating the causes (models) from known results (data). In the context of sports, a forward problem includes generating or predicting outcomes based on specific tactics or formations. In research using simulations, it is common to simulate results based on predefined behaviors or tactics, which is referred to as forward analysis. On the other hand, an inverse problem involves analyzing the causes or underlying tactics from actual match results or data. This is a common practice for human coaches and players during post-match analysis, which is known as inverse analysis.

Inverse planning methods derive the action model or reward function from observed data using statistical learning techniques. This approach also leverages RL frameworks using real-world data in a physical space. While it shares some similarities with predictive modeling, the main focus of inverse planning is to evaluate actions and states to achieve specific goals, rather than extracting features or predicting trajectories. Forward planning methods, on the other hand, assume optimal criteria for an agent's movements based on a predefined reward function, such as a score in team sports. This contrasts with pattern-based methods, which often rely on historical data to identify patterns and predict outcomes without explicitly planning for long-term goals. However, when dealing with simulation agents that behave intelligently like humans or problems that involve replicating movements from human behavior data, both forward and inverse analysis may be required, making it difficult to simplify the approach. For instance, estimating tactics during simulations is an inverse problem, while trajectory prediction in human behavior analysis is a forward problem. The next sections introduces inverse approaches for player and team evaluations and forward approaches for learning-based simulation.

4.3 Player and Team Evaluations as Inverse Approach

In the inverse approach from real-world data, the key challenges are to evaluate states and actions, and to replicate actual behaviors. Due to the difficulty in complete modeling, researchers have often focused on sub-problems such as action evaluation and policy/reward estimation. In specific team plays, deep RL to estimate the quality of the defensive actions was used in ball-screen defense in basketball [64]. To value on-ball actions, several studies have estimated state-action value function (Q-function) or other policy functions in soccer event data [29, 62, 63] and with tracking data [41, 43, 44], ice hockey [30, 48], and both [54], badminton [9], and basketball [6, 65], which can be considered as offline RL (reviewed by [24]). However, they often

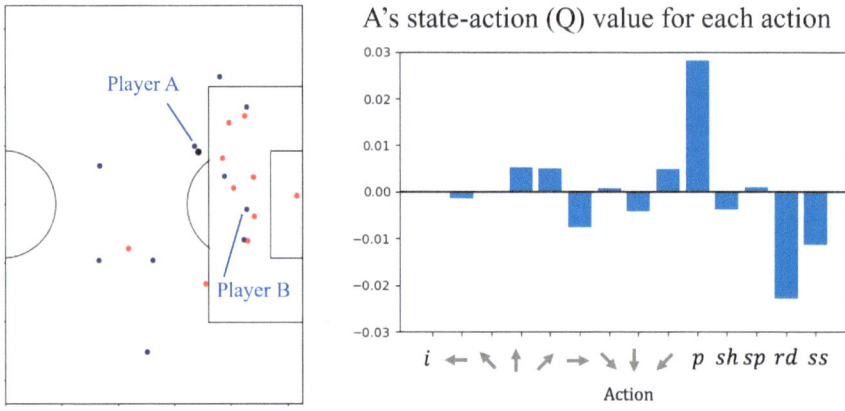

Fig. 4.2 Example of estimated Q-values in the RL model from actual soccer data [38]. **a** shows an example attack. Blue, red, and black indicate the attacker, defender, and the ball, respectively (note that the ball is over a defender). **b** show the Q-value of player A for each action using the RL model with action supervision. It takes the highest value in a pass action, indicating that the pass action is suggested to be the best option in this situation. Note that i, p, sh, sp, rd, ss correspond to idle, pass, shot, sprint, decelerate, and sprint end, respectively, and arrows correspond to the direction of movement

modeled team representation with on-ball events and did not evaluate off-ball players in all time steps. Our group [38] considered a holistic RL model estimating multiple players' Q-functions for simultaneously evaluating on- and off-ball players even when no event occurs as illustrated in Fig. 4.2.

In terms of inverse RL (IRL), research on estimating reward functions has also been conducted [33, 42]. To estimate policy functions, researchers have sometimes performed trajectory prediction through imitation learning [13, 20, 21, 58] and behavioral modeling [14, 27, 68, 71], aiming to mimic (rather than optimize) a policy using neural networks. For other examples using MDP, researchers have modeled the transition probabilities and shot policy tensors and have simulated seasons under alternative shot policies of interest [46]. A sequential team selection model was also considered in soccer to optimize long-term team performance by mitigating injury risks [12].

4.4   Learning-Based Simulation as Forward Approach

Learning-based simulation can basically utilize RL but can be categorized in various ways based on the nature of the data and learning methodologies. One of the primary categories is RL from scratch (e.g., without expert demonstrations or real-world data). This involves agents learning optimal policies purely through interaction with the environment, often starting from scratch without any prior data. These agents rely heavily on trial and error to maximize rewards, exploring different actions and learning from the results. This approach is particularly useful in scenarios where prior data is unavailable or where the environment is too complex to model explicitly before the learning process begins.

Another category is imitation learning without RL, where agents learn by mimicking the behavior observed in demonstrations provided by experts, rather than through rewards and penalties (described in Sect. 3.4.4). Additionally, learning from demonstration (with expert data and online RL) [47] combines elements of both RL and imitation learning. In this approach, agents start by learning from demonstration data and then refine their policies through RL techniques. This approach can include IRL with online RL, where agents infer the reward function that a demonstrator is optimizing, and then use this inferred reward function to learn an optimal policy. This method is beneficial when the reward function is not explicitly provided.

Based on the above categories, in team sports, RL from scratch has been intensively studied to improve learning efficiency, computation, and communication in GFootball (e.g., [11, 26, 31, 45]) and Fever Basketball Defense [57]. Meanwhile, in the RoboCup 2D Simulation League, methods that do not use RL but instead combine rule-based and machine learning are currently superior (e.g., [1]). With Simple Team Sports Simulator, the process by which agents autonomously discover high-level skills needed for collaborative task completion was investigated [66, 73], focusing on human-AI cooperation.

However, the above studies did not utilize real-world data for realistic sports simulators. Before creating simulators, we compared ball-passing behaviors in artificial agents using GFootball with those of professional soccer players, though a gap still exists between forward and backward approaches [49]. The gap between forward and backward approaches will be discussed in Sects. 4.5 and 5.1. Recently we integrated both methodologies to balance the reproducibility of imitation and the generalization needed to obtain rewards [15]. We used chase-and-escape and soccer tasks with the different dynamics between the unknown source and target environments as illustrated in Figs. 4.3 and 4.4, respectively. We show that our approach for both tasks achieved a balance between the reproducibility and the generalization ability compared with the baselines. In the soccer task (see Fig. 4.4), although our approach correctly learned Q-function values in which the higher values were observed in pass and shot actions for the passer (agent 1) and shooter (agent 0) compared with DQN, respectively, our approach did not reproduce the demonstration movements toward the goal. Combining to learn both on- and off-ball actions is a future work. For example, creating a better multi-agent simulator and RL model utilizing domain knowledge is a possible future research direction for reproducing not only actions but also movements (e.g., [59]).

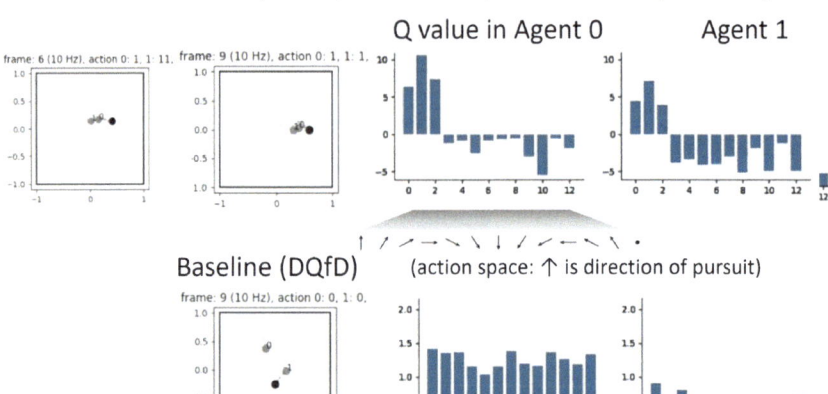

Fig. 4.3 Example RL results in the 2vs1 chase-and-escape task with our approach [15]. Our approach (DQAAS: Deep Q-learning with adaptive action supervision, right top) and the baseline (DQfD [16], center), and the demonstration in the source domain (left) are shown. Note that the two predators in the target domain are slower than those in the source domain, thus it is difficult to learn the correct actions and a domain adaptation approach is required. We adopt an approach that combines RL and supervised learning by selecting actions of demonstrations in RL based on the minimum distance of dynamic time warping for utilizing the information of the unknown source dynamics. Histograms are the Q-function values for each action. There are 13 actions including acceleration in 12 directions every 30 degrees in the relative coordinate system (action 0 means moving towards the prey) and doing nothing (action 12: round point). The histograms show that our approach learned the correct action value regarding the direction of pursuit, but the baseline approach did not

4.5   Future Research Topics

Future research in the field of learning-based sports simulations should first focus on enhancing the definitions and representations of state, action, and reward. Currently, states are often defined too simplistically, using only positions and velocities (e.g., [15, 38]). However, real players make decisions based on a more complex understanding, often involving tactical knowledge and selective attention to certain information. Therefore, integrating these aspects such as by combining them with mathematical models used in space evaluation could provide a more realistic and effective state representation. Similarly, more sophisticated models can be created by refining action and reward definitions such as in [43, 44], which accurately reflect the intricate dynamics of team sports.

When using data, the combination of states and actions creates a vast array of possibilities that go beyond simple pattern-based prediction problems. This means that more sophisticated methods are required to use more data than pattern-based approaches and/or to handle this complexity. Mathematical models can potentially reduce the reliance on extensive data by providing a structured way to model player

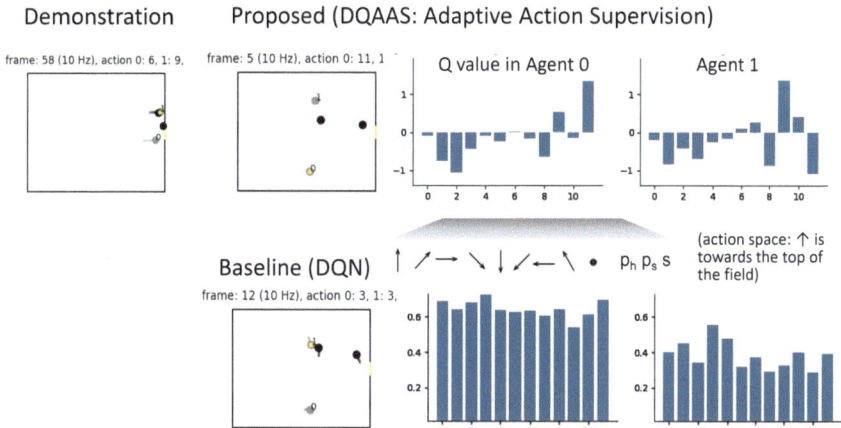

Fig. 4.4 Example RL results in the 2vs2 soccer task with our approach [15]. Configurations are the same as Fig. 4.3. In the demonstration and DQAAS, the agents obtained the goal, but the DQN failed. There are 12 actions including the movement in 8 directions (with constant velocity) every 45 degrees in the relative coordinate system (actions 0-7 and action 0 means moving toward the center of the goal), doing nothing (action 8: round point), and high pass (action 9: p_h), short pass (action 10: p_s), and shot (action 11: s), which are partially based on GFootball [18]. The histograms show that our approach learned the correct action value regarding the shot (agent 0) and pass (agent 1), but the baseline approach did not

behaviors. Moreover, current research has a significant gap between forward and backward approaches. The inverse approach aims to replicate actual behavior from real-world data, but it is uncertain whether the actions generated are genuinely reward-earning since forward simulation is not performed. On the other hand, the forward approach generates reward-earning actions in a virtual space without measured data, yet their real-world applicability remains uncertain. A hybrid approach such as in [15] that combines these methods could address the challenges inherent in both and lead to more robust and accurate models.

The integration of game theory and RL also presents a promising but largely unexplored avenue. By combining the strategic interaction models of game theory with the adaptive learning capabilities of RL, it is possible to develop agents that can optimize their behavior not only in response to the environment but also in response to the strategies of other agents. This could significantly enhance the realism and effectiveness of simulations in team sports analytics.

4.6 Summary

This chapter explored the potential of learning-based agent models in sports analytics, highlighting their ability to simulate complex scenarios, evaluate tactical choices, and suggest optimal actions. We discussed the fundamentals of agent modeling, including the MDP, and examined various simulation platforms such as RoboCup and GFootball. This chapter differentiated from traditional rule-based simulations, emphasizing the advantages of learning-based approaches. We also discussed the application of game theory and the roles of inverse and forward planning methods in evaluating actions and generating realistic simulations. In the next chapter, future perspectives and ecosystems, covering advanced research directions, practical applications, and the shaping of future ecosystems will be introduced.

References

1. Akiyama, H., Nakashima, T., Hatakeyama, K., Fujikawa, T., Hishiki, A.: Helios2023: Robocup 2023 soccer simulation 2d competition champion. In: Robot World Cup, pp. 386–394. Springer (2023)
2. Alguacil, F.P., Arce, P.P., Sumpter, D., Fernandez, J.: Seeing in to the future: using self-propelled particle models to aid player decision-making in soccer. In: Proceedings of the MIT Sloan Sports Analytics Conference (2020)
3. Arulkumaran, K., Deisenroth, M.P., Brundage, M., Bharath, A.A.: Deep reinforcement learning: a brief survey. IEEE Signal Proc. Mag. **34**(6), 26–38 (2017)
4. Azar, O.H., Bar-Eli, M.: Do soccer players play the mixed-strategy nash equilibrium? Appl. Econ. **43**(25), 3591–3601 (2011)
5. Buzzacchi, L., Pedrini, S.: Does player specialization predict player actions? Evidence from penalty kicks at fifa world cup and uefa euro cup. Appl. Econ. **46**(10), 1067–1080 (2014)
6. Chen, X., Jiang, J.-Y., Jin, K., Zhou, Y., Liu, M., Brantingham, P.J., Wang, W.: Reliable: offline reinforcement learning for tactical strategies in professional basketball games. In: Proceedings of the 31st ACM International Conference on Information & Knowledge Management, pp. 3023–3032 (2022)
7. Chiappori, P.-A., Levitt, S., Groseclose, T.: Testing mixed-strategy equilibria when players are heterogeneous: the case of penalty kicks in soccer. Am. Econ. Rev. **92**(4), 1138–1151 (2002)
8. Coloma, G.: The penalty-kick game under incomplete information. University of CEMA Economics Serie Documentos de Trabajo (487) (2012)
9. Ding, N., Takeda, K., Fujii, K.: Deep reinforcement learning in a racket sport for player evaluation with technical and tactical contexts. IEEE Access **10**, 54764–54772 (2022)
10. Emara, N., Owens, D.M., Smith, J., Wilmer, L.: Minimax on the gridiron: serial correlation and its effects on outcomes in the national football league (2014). SSRN 2502193
11. Espeholt, L., Marinier, R., Stanczyk, P., Wang, K., Michalski, M.: SEED RL: scalable and efficient deep-rl with accelerated central inference. In: International Conference on Learning Representations (2019)
12. Everett, G., Beal, R., Matthews, T., Norman, T.J., Ramchurn, S.D.: The strain of success: a predictive model for injury risk mitigation and team success in soccer. In: Proceedings of the MIT Sloan Sports Analytics Conference (2024)
13. Fujii, K., Takeishi, N., Kawahara, Y., Takeda, K.: Decentralized policy learning with partial observation and mechanical constraints for multiperson modeling. Neural Netw. **171**, 40–52 (2024)
14. Fujii, K., Takeuchi, K., Kuribayashi, A., Takeishi, N., Kawahara, Y., Takeda, K.: Estimating counterfactual treatment outcomes over time in multi-vehicle simulation. In: Proceedings of the 30th International Conference on Advances in Geographic Information Systems (SIGSPATIAL'22), pp. 1–4 (2022)
15. Fujii, K., Tsutsui, K., Scott, A., Nakahara, H., Takeishi, N., Kawahara, Y.: Adaptive action supervision in reinforcement learning from real-world multi-agent demonstrations. In: 16th

International Conference on Agents and Artificial Intelligence (ICAART'24), vol. 2, pp. 27–39 (2024)

16. Hester, T., Vecerik, M., Pietquin, O., Lanctot, M., Schaul, T., Piot, B., Horgan, D., Quan, J., Sendonaris, A., Osband, I., et al.: Deep q-learning from demonstrations. In: Proceedings of the Thirty-Second AAAI Conference on Artificial Intelligence and Thirtieth Innovative Applications of Artificial Intelligence Conference, pp. 3223–3230 (2018)

17. Kitano, H., Asada, M., Kuniyoshi, Y., Noda, I., Osawa, E.: Robocup: the robot world cup initiative. In: Proceedings of the First International Conference on Autonomous Agents, pp. 340–347 (1997)

18. Kurach, K., Raichuk, A., Stańczyk, P., Zając, M., Bachem, O., Espeholt, L., Riquelme, C., Vincent, D., Michalski, M., Bousquet, O., et al.: Google research football: A novel reinforcement learning environment. In: Proceedings of the AAAI Conference on Artificial Intelligence, vol. 34, pp. 4501–4510 (2020)

19. Lanctot, M., Zambaldi, V., Gruslys, A., Lazaridou, A., Tuyls, K., Pérolat, J., Silver, D., Graepel, T.: A unified game-theoretic approach to multiagent reinforcement learning. Adv. Neural. Inf. Process. Syst. **30**, 4193–4206 (2017)

20. Le, H.M., Carr, P., Yue, Y., Lucey, P.: Data-driven ghosting using deep imitation learning. In: Proceedings of MIT Sloan Sports Analytics Conference (2017)

21. Le, H.M., Yue, Y., Carr, P., Lucey, P.: Coordinated multi-agent imitation learning. In: Proceedings of the 34th International Conference on Machine Learning, Vol. 70, pp. 1995–2003 (2017). (JMLR. org, 2017)

22. Leduc, A., Cioppa, A., Giancola, S., Ghanem, B., Van Droogenbroeck, M.: Soccernet-depth: a scalable dataset for monocular depth estimation in sports videos. In: Proceedings of the IEEE/CVF Conference on Computer Vision and Pattern Recognition, pp. 3280–3292 (2024)

23. Lennartsson, J., Lidstrom, N., Lindberg, C.: Game intelligence in team sports. PLoS ONE **10**, e0125453 (2015)

24. Levine, S., Kumar, A., Tucker, G., Fu, J.: Offline reinforcement learning: tutorial, review, and perspectives on open problems (2020). arXiv:2005.01643

25. Levitt, S., Chiappori, P., Groseclose, T.: Testing mixed-strategy equilibria when players are heterogeneous: the case of penalty kicks in soccer. Am. Econ. Rev. **92**, 1138–1151 (2002)

26. Li, C., Wang, T., Chengjie, W., Zhao, Q., Yang, J., Zhang, C.: Celebrating diversity in shared multi-agent reinforcement learning. Adv. Neural. Inf. Process. Syst. **34**, 3991–4002 (2021)

27. Li, L., Yao, J., Wenliang, L., He, T., Xiao, T., Yan, J., Wipf, D., Zhang, Z.: GRIN: Generative relation and intention network for multi-agent trajectory prediction. Adv. Neural. Inf. Process. Syst. **34**, 27107–27118 (2021)

28. Lin, K.: Applying game theory to volleyball strategy. Int. J. Perform. Anal. Sport **14**(3), 761–774 (2014)

29. Liu, G., Luo, Y., Schulte, O., Kharrat, T.: Deep soccer analytics: learning an action-value function for evaluating soccer players. Data Min. Knowl. Disc. **34**(5), 1531–1559 (2020)

30. Liu, G., Schulte, O.: Deep reinforcement learning in ice hockey for context-aware player evaluation. In: Proceedings of the 27th International Joint Conference on Artificial Intelligence, pp. 3442–3448 (2018)

31. Liu, I.-J., Ren, Z., Yeh, R.A., Schwing, A.G.: Semantic tracklets: an object-centric representation for visual multi-agent reinforcement learning. In: 2021 IEEE/RSJ International Conference on Intelligent Robots and Systems (IROS), pp. 5603–5610. IEEE (2021)

32. Lowe, R., Wu, Y.I., Tamar, A., Harb, J., Pieter Abbeel, O.A.I., Mordatch, I.: Multi-agent actor-critic for mixed cooperative-competitive environments. Adv. Neural Inf. Process. Syst. **30**, 6382–6393 (2017)

33. Luo, Y., Schulte, O., Poupart, P.: Inverse reinforcement learning for team sports: valuing actions and players. In: Bessiere, C. (ed.) Proceedings of the Twenty-Ninth International Joint Conference on Artificial Intelligence, IJCAI-20. International Joint Conferences on Artificial Intelligence Organization, vol. 7, pp. 3356–3363 (2020)

34. Mao, Y., Wu, C., Chen, X., Hu, H., Jiang, J., Zhou, T., Lv, T., Fan, C., Hu, Z., Wu, Y., et al.: Stylized offline reinforcement learning: extracting diverse high-quality behaviors from heterogeneous datasets. In: The Twelfth International Conference on Learning Representations (2024)
35. Mendes-Neves, T., Mendes-Moreira, J., Rossetti, R.J.F.: A data-driven simulator for assessing decision-making in soccer. In: EPIA Conference on Artificial Intelligence, pp. 687–698. Springer (2021)
36. Mnih, V., Kavukcuoglu, K., Silver, D., Rusu, A.A., Veness, J., Bellemare, M.G., Graves, A., Riedmiller, M., Fidjeland, A.K., Ostrovski, G., et al.: Human-level control through deep reinforcement learning. Nature **518**(7540), 529–533 (2015)
37. Moschini, G.C.: Nash equilibrium in strictly competitive games: live play in soccer. Econ. Lett. **85**(3), 365–371 (2004)
38. Nakahara, H., Tsutsui, K., Takeda, K., Fujii, K.: Action valuation of on-and off-ball soccer players based on multi-agent deep reinforcement learning. IEEE Access **11**, 131237–131244 (2023)
39. Nash, J.: Non-cooperative games. Ann. Math. 286–295 (1951)
40. Palacios-Huerta, I.: Professionals play minimax. Rev. Econ. Stud. **70**(2), 395–415 (2003)
41. Rahimian, P., Oroojlooy, A., Toka, L.: Towards optimized actions in critical situations of soccer games with deep reinforcement learning. In: 2021 IEEE 8th International Conference on Data Science and Advanced Analytics (DSAA), pp. 1–12. IEEE (2021)
42. Rahimian, P., Toka, L.: Inferring the strategy of offensive and defensive play in soccer with inverse reinforcement learning. In: Machine Learning and Data Mining for Sports Analytics, pp. 26–38. Springer (2022)
43. Rahimian, P., Van Haaren, J., Toka, L.: Towards maximizing expected possession outcome in soccer. Int. J. Sports Sci. Coach. **19**(1), 230–244 (2024)
44. Rahimian, P., Van Haaren, J., Abzhanova-Laszlo Toka, T.: Beyond action valuation: a deep reinforcement learning framework for optimizing player decisions in soccer. In: 16th Annual MIT Sloan Sports Analytics Conference (2022)
45. Roy, J., Barde, P., Harvey, F., Nowrouzezahrai, D., Pal, C.: Promoting coordination through policy regularization in multi-agent deep reinforcement learning. Adv. Neural. Inf. Process. Syst. **33**, 15774–15785 (2020)
46. Sandholtz N., Bornn, L.: Replaying the NBA. In: Proceedings of the MIT Sloan Sports Analytics Conference (2018)
47. Schaal, S.: Learning from demonstration. Adv. Neural. Inf. Process. Syst. **9**, 1040–1046 (1996)
48. Schulte, O., Khademi, M., Gholami, S., Zhao, Z., Javan, M., Desaulniers, P.: A markov game model for valuing actions, locations, and team performance in ice hockey. Data Min. Knowl. Disc. **31**(6), 1735–1757 (2017)
49. Scott, A., Fujii, K., Onishi, M.: How does AI play football? An analysis of RL and real-world football strategies. In: 14th International Conference on Agents and Artificial Intelligence (ICAART'22), vol. 1, pp. 42–52 (2022)
50. Sindik, J., Vidak, N.: Application of game theory in describing efficacy of decision making in sportsman's tactical performance in team sports. Interdiscip. Descr. Complex Syst. Sci. J. **6**, 53–66 (2008)
51. Skinner, B.: The price of anarchy in basketball. J. Quant. Anal. Sports **6**(1) (2010)
52. Somers, C., Rupert, J., Zhao, Y., Borovikov, I., Yang, J., Beirami, A.: Simple Team Sports Simulator (STS2) (February 2020)
53. Spearman, W., Basye, A., Dick, G., Hotovy, R., Pop, P.: Physics-based modeling of pass probabilities in soccer. In: Proceeding of the 11th MIT Sloan Sports Analytics Conference (2017)
54. Sun, X., Davis, J., Schulte, O., Liu, G.: Cracking the black box: distilling deep sports analytics. In: Proceedings of the 26th ACM SIGKDD Conference on Knowledge Discovery and Data Mining, pp. 3154–3162 (2020)
55. Sutton, R.S., Barto, A.G.: Reinforcement Learning: An Introduction. MIT Press (2018)
56. Tadelis, S.: Game Theory: An Introduction. Princeton University Press (2013)

57. Tang, H., Hao, J., Lv, T., Chen, Y., Zhang, Z., Jia, H., Ren, C., Zheng, Y., Meng, Z., Fan, C., et al.: Hierarchical deep multiagent reinforcement learning with temporal abstraction (2018). arXiv:1809.09332

58. Teranishi, M., Fujii, K., Takeda, K.: Trajectory prediction with imitation learning reflecting defensive evaluation in team sports. In: 2020 IEEE 9th Global Conference on Consumer Electronics (GCCE), pp. 124–125. IEEE (2020)

59. Tsutsui, K., Takeda, K., Fujii, K.: Synergizing deep reinforcement learning and biological pursuit behavioral rule for robust and interpretable navigation. In: 1st Workshop on the Synergy of Scientific and Machine Learning Modeling in International Conference on Machine Learning (2023)

60. Tsutsui, K., Tanaka, R., Takeda, K., Fujii, K.: Collaborative hunting in artificial agents with deep reinforcement learning. eLife **13**, e85694 (2024)

61. Tuyls, K., Omidshafiei, S., Muller, P., Wang, Z., Connor, J., Hennes, D., Graham, I., Spearman, W., Waskett, T., Steel, D., et al.: Game plan: What ai can do for football, and what football can do for ai. J. Artif. Intell. Res. **71**, 41–88 (2021)

62. Van Roy, M., Robberechts, P., Yang, W.-C., De Raedt, L., Davis, J.: Leaving goals on the pitch: evaluating decision making in soccer. In: Proceedings of the MIT Sloan Sports Analytics Conference (2021)

63. Van Roy, M., Robberechts, P., Yang, W.-C., De Raedt, L., Davis, J.: A Markov framework for learning and reasoning about strategies in professional soccer. J. Artif. Intell. Res. **77**, 517–562 (2023)

64. Wang, X., Girshick, R., Gupta, A., He, K.: Non-local neural networks. In: Proceedings of the IEEE Conference on Computer Vision and Pattern Recognition, pp. 7794–7803 (2018)

65. Yanai, C., Solomon, A., Katz, G., Shapira, B., Rokach, L.: Q-ball: Modeling basketball games using deep reinforcement learning. In: Proceedings of the AAAI Conference on Artificial Intelligence , vol. 36, pp. 8806–8813 (2022)

66. Yang, J., Borovikov, I., Zha, H.: Hierarchical cooperative multi-agent reinforcement learning with skill discovery. In: Proceedings of the 19th International Conference on Autonomous Agents and Multi Agent Systems, pp. 1566–1574 (2020)

67. Yee, A., Rodríguez, R., Alvarado, M.: Analysis of strategies in American football using nash equilibrium. In: Artificial Intelligence: Methodology, Systems, and Applications: 16th International Conference, AIMSA 2014, pp. 286–294. Springer (2014)

68. Yeh, R.A., Schwing, A.G., Huang, J., Murphy, K.: Diverse generation for multi-agent sports games. In: The IEEE Conference on Computer Vision and Pattern Recognition (CVPR), pp. 4610–4619 (2019)

69. Yeung, C., Fujii, K.: A strategic framework for optimal decisions in football 1-vs-1 shot-taking situations: an integrated approach of machine learning, theory-based modeling, and game theory. Complex Intell. Syst. 1–20 (2024)

70. Yokoyama, K., Shima, H., Fujii, K., Tabuchi, N., Yamamoto, Y.: Social forces for team coordination in ball possession game. Phys. Rev. E **97**(2), 022410 (2018)

71. Zhan, E., Zheng, S., Yue, Y., Sha, L., Lucey, P.: Generating multi-agent trajectories using programmatic weak supervision. In: International Conference on Learning Representations (2019)

72. Zhang, K., Yang, Z., Başar, T.: Multi-agent reinforcement learning: a selective overview of theories and algorithms. In: Handbook of Reinforcement Learning and Control, pp. 321–384 (2021)

73. Zhao, Y., Borovikov, I., Rupert, J., Somers, C., Beirami, A.: On multi-agent learning in team sports games (2019). arXiv:1906.10124

74. Zhu, L., Rematas, K., Curless, B., Seitz, S., Kemelmacher-Shlizerman, I.: Reconstructing NBA players. In: Proceedings of the European Conference on Computer Vision (ECCV) (Aug 2020)

Chapter 5
Future Perspectives and Ecosystems

Abstract In this final chapter, the necessity of integrating computer vision, predictive analysis, and learning-based agent modeling in sports analytics is explored to address the complex and dynamic nature of sports movements. A hypothesis on advanced research directions is presented, emphasizing the integration of real-world data and digital modeling. This integration enables more comprehensive systems capable of prediction, play evaluation, and optimal play suggestions. Furthermore, the practical deployment of these technologies in real-world scenarios is discussed, focusing on their impact on various levels of sports. Finally, the formation of future ecosystems that support these advancements is explored, highlighting the importance of open approaches, standardization, and collaboration.

Keywords Digital modeling · Performance monitoring · Strategic decision-making · Fan engagement · Open-source

5.1 Introduction

In Chaps. 2–4, computer vision, predictive analysis, and learning-based agent modeling are explored but are currently being researched independently in terms of sports analytics. While each of these technologies has demonstrated significant potential on its own, their real-world applications necessitate an integrated approach. To fully leverage the benefits of these advancements, it is important to combine these technologies, creating comprehensive systems that can address the complex and dynamic nature of sports movements.

This final chapter first introduces a hypothesis of advanced research directions, highlighting the integration of real-world data and digital modeling that are shaping the future of sports analytics in Sect. 5.2. Following this, we discuss the deployment of these technologies in real-world scenarios, focusing on their practical implementations and the impact they can have on sports at various levels in Sect. 5.3. Finally, the formation of future ecosystems necessary to support and sustain these advancements is explored in Sect. 5.4, emphasizing the importance of an open approach, collaboration, standardization, and other considerations.

© The Author(s) 2025
K. Fujii, *Machine Learning in Sports*,
SpringerBriefs in Computer Science, https://doi.org/10.1007/978-981-96-1445-5_5

It is important to note that this chapter contains several hypotheses about future directions, drawn from the author's personal experiences and insights gained through non-published communications. While these ideas are forward-looking, they may not yet be empirically validated. Readers are encouraged to consider these points as exploratory rather than definitive conclusions. Additionally, while the term "machine learning" is often used in this book, the more broadly recognized term "AI" (artificial intelligence) can be used interchangeably in certain contexts for better accessibility in this chapter. AI encompasses a broader field, of which machine learning is a part, making it more familiar to a wider audience.

5.2   A Hypothesis of Advanced Research Directions

While advancements in individual technologies for learning-based sports analytics are crucial, the true potential lies in the integration of real-world data and digital modeling. This integration is important because it enables the creation of accurate digital representations of biological entities, allowing for comprehensive analysis, prediction, evaluation, and suggestion of optimal tactics and strategies. By combining data from cameras, wearable sensors, and other monitoring technologies with advanced machine learning algorithms and simulation models, we can achieve a deeper understanding of player and team performances. This holistic approach not only has the potential to improve the effectiveness of play evaluations but also may facilitate feedback and enhancements, which could lead to significant improvements in training, conditioning, and game performance. Here, this hypothesis is explored in detail, as illustrated in Fig. 5.1, examining how these potential enhancements can be achieved and their implications for the future of sports analytics.

To illustrate this, Fig. 5.1 presents a comprehensive framework for integrated learning-based sports analytics. The diagram outlines the four main processes: (1) data acquisition, (2) modeling, (3) simulation, and (4) feedback and enhancement. To realize this, there are three major challenges within these processes: (a) automation of data acquisition, (b) addressing the domain gap between real-world and simulated environments, and (c) effective implementation of learned models in practical applications. In particular, first, (b) the critical step of real-to-sim domain adaptation is considered in Sect. 5.2.1, where real-world data in a physical space is adapted to simulation in a digital space. Other challenges for each component are explained in Sect. 5.2.2. Finally, long future prospects and benefits of digital modeling in sports are explored in Sect. 5.2.3.

5.2.1   Integration of Real-World Data and Digital Modeling

When modeling real-world biological multi-agents, domain gaps may occur between behaviors in the sources (real-world data) with unknown dynamics and targets (sim-

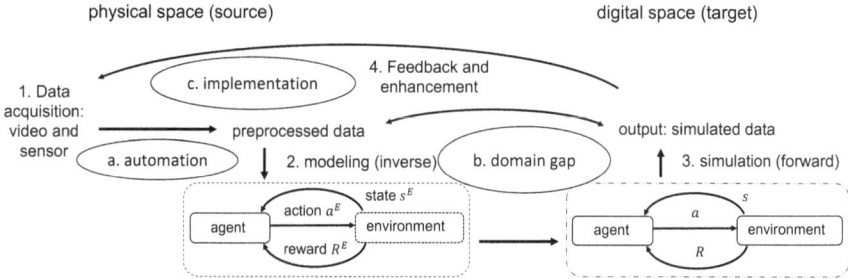

Fig. 5.1 A hypothesis in integrated learning-based sports analytics framework. This diagram illustrates the four main processes in learning-based simulation: (1) data acquisition, (2) modeling, (3) simulation, and (4) feedback and enhancement. From (2) modeling to (3) simulation, a real-to-sim domain adaptation problem should be considered, in which the source and target are real-world data in physical space and simulated data in digital space, respectively. It also highlights three significant challenges within these processes: **a** automation of data acquisition, **b** addressing the domain gap between real-world and simulated environments, and (c) effective implementation of learned models in practical applications

ulation in digital space). The opposite configuration, known as sim-to-real [28], transfers knowledge from simulations or human demonstrations to real-world applications, such as robotics [20, 30]. However, domain adaptation in real-world situations often deals with unknown source dynamics, making it difficult to utilize explicit transition models. In contrast, real-to-sim domain adaptation focuses on translating unknown real-world data into simulated environments (Fig. 5.1). In team sports, for example, this means leveraging reinforcement learning (RL) in digital space for flexible adaptation to complex environments [21, 24] and data-driven modeling for reproducing real-world behaviors [12, 22, 44]. Despite the gap between these forward and backward approaches in multi-agent RL (MARL) scenarios, an integrated approach that combines both strengths is essential to bridge this divide and fully harness the potential of sports analytics.

When considering the advancement of learning-based sports analytics, the integration of real-world data with digital models is expected to be an important area of development. This integration is key to enhancing the utility of prediction, play evaluation, and optimal play suggestion technologies. By comparing this integration to well-known two concepts, the importance of integration in transforming sports analytics can be further emphasized. The first concept, *cyber-physical systems* (CPS), which tightly integrates physical processes with digital computations and networking (e.g., [33]). CPS monitors and controls physical processes with feedback loops that allow physical actions to influence computations and vice versa. In sports, CPS can be utilized to monitor athlete performance, refine training routines, and enhance game tactics and strategies in real-time. However, implementing CPS in sports analytics faces significant challenges. The real-time data processing and complex integration required are currently limited by technological constraints, such as the need for high-speed networking infrastructure, real-time and accurate data processing capabilities, and the seamless integration of diverse data sources. These requirements are often difficult to meet in dynamic and variable environments like sports fields.

Meanwhile, *digital twins* offers a good example for creating virtual replicas of physical/biological entities [13] (in this case, athletes or sports environments). Digital twins use real-time data to mirror the physical world in a digital environment, facilitating advanced simulations, analysis, and enhancement. They enable continuous monitoring and provide insights that can improve performance and predict future outcomes. However, the creation and maintenance of accurate digital twins require high-fidelity data and sophisticated modeling techniques (usually examined in physical entities, rather than biological ones), which can be resource-intensive and complex to implement in sports analytics. The current state of technology in sports analytics often lacks the necessary precision and resolution in data acquisition, as well as the computational power required for real-time updates and detailed simulations. For these reasons, focusing on the integration of real-world data and digital modeling is more appropriate at this stage. This approach also leverages real-world data to create accurate digital models for prediction and tactical analysis, ultimately bridging the gap between physical and digital spaces in sports analytics.

Integrating real-world data with digital models involves four steps as illustrated in Fig. 5.1:

- Data Acquisition: Real-world data is collected from various sources, including cameras, sensors, and other manual/automatic monitoring technologies. This data includes field, tracking, event, pose, and other information as described in Chaps. 1 and 2.
- Modeling: Digital models are created to replicate sports behaviors. These models use the collected data to accurately reflect the current state of the athletes and sports environments. The process involves using machine learning algorithms and other techniques to develop models that can predict, evaluate, and suggest plays based on the input data as described in Chaps. 3 and 4.
- Simulation: Once the digital models are established, simulations can be run to test different scenarios and strategies. These simulations help in understanding how various factors interact and influence the outcomes, allowing for experimentation without impacting the real-world environment, as described in Chap. 4.
- Feedback and Enhancement: Insights from the digital models and simulations are used to make adjustments to the real-world environments. For example, training programs and game strategies can be refined. Necessary components to realize them and detailed examples are described in Sects. 5.2.2 and 5.2.3, respectively.

This approach leverages real-world data to create accurate digital models, effectively bridging the gap between physical space and digital space. Next, future developments in each component of digital modeling in sports are discussed.

5.2.2   Future Developments in Each Component for Digital Modeling

The integration of real-world data and digital modeling in sports depends on advancements in several key components. These components collectively enhance the capability to monitor, model, simulate, and enhance various aspects of athletic performance and sports strategies. This section explores the progress in data acquisition, modeling, simulation, and feedback and enhancement, which are crucial for developing future intelligent digital models in sports. In this context, the definition of "real-time" refers to data or information being processed and utilized almost immediately after it is generated. This typically implies processing within a delay of a few milliseconds to a few seconds. However, in practice, achieving strict real-time processing can be technically challenging. Here, the term "near real-time" is used, which allows for delays of a few minutes to a few hours.

Data Acquisition

The current state of annotated data for sports analytics is overwhelmingly insufficient in both quantity and diversity. The only somewhat sufficient source of annotated data is video footage of soccer games, largely thanks to the efforts of the SoccerNet community. However, for other sports and different filming methods, such as those in [31], the amount of available annotated data is still severely lacking. While data acquisition approaches are discussed in more detail in Sects. 5.3 and 5.4, this section focuses on the technical approaches. From a technical standpoint, it is challenging to fully automate the annotation process, so strategies need to be developed to achieve complete data sets with minimal manual annotation. This could involve techniques like human-in-the-loop or active learning (e.g., action spotting in soccer [14] and action recognition in basketball [1]), which incorporates human feedback during the learning process, semi-supervised learning that combines a small amount of labeled data with a large amount of unlabeled data (e.g., predicting trajectories [11], extract tactical patterns in soccer [4], and screen-play detection in basketball [45]), and transfer learning which leverages data from other leagues (e.g., [6]) or even sports to improve the model's performance.

Another promising area for technical advancement is the end-to-end approach using neural networks. While component-specific methods such as tracking (e.g., [8]) and event detection (e.g., [18]) are already in use, there is potential for more comprehensive solutions, such as directly estimating player positions on the field or court from video footage. As the volume of data increases, these methods are expected to become more efficient and faster. Additionally, the development of methods with multiple modalities, which integrate sensor data, pose data, language, and sound as discussed in Chap. 2, is essential for creating more accurate digital models. These advancements will significantly enhance the accuracy and efficiency of subsequent modeling processes.

Modeling

If there is sufficient data in both quality and quantity, it is possible to create digital models of players using purely data-driven approaches. For instance, in outcome prediction, event prediction, and trajectory prediction, transformer-based methods that follow scaling laws (e.g., [19]) can achieve accurate predictions if the data quality and its quantity are sufficient. However, when it comes to agent-based models, as discussed in Chap. 4, it is extremely challenging to gather enough data to cover all possible combinations of unobserved states and actions. Therefore, incorporating some form of domain knowledge is necessary to enhance the modeling process.

One of the most feasible approaches is the introduction of mathematical models to evaluate spatial configurations, which, once computational complexity is addressed, can become quite effective. Additionally, by defining fundamentally important and simplified states and actions, it is possible to reduce the complexity of the state and action spaces, thereby increasing the coverage of these spaces by the available data (expressed in more precise reinforcement learning terminology). It is also important to consider the strategic interactions between players, necessitating the integration of game theory into the models. To effectively incorporate such agent decision making, dynamic game theory will be promising for future research (e.g., [3]). This approach is necessary to account for the evolving strategies and decisions of players in near real-time, thus providing a more robust and realistic model of player behavior and interactions.

Simulation

Developing realistic model simulations using real-world data can enable the testing and validation of various tactics and strategies. A central issue, as mentioned in Sect. 5.2.1, is bridging the gap between inverse modeling from real-world data and forward simulation modeling. Human behavior may be a mix of deductive rules based on knowledge and inductive actions based on experience. Therefore, it is challenging to achieve realistic simulations using purely rule-based or purely learning-based approaches. To address this, a hybrid methodology is necessary, combining rule-based systems that capture skilled movements with learning-based methods that allow for flexible adaptation to different situations (e.g., [38]). This hybrid approach would enable the creation of agent models that can both follow known strategies and adapt to new scenarios dynamically.

Another advanced topic is the development of pose simulators using reinforcement learning, as demonstrated by DeepMind's MuJoCo (Multi-Joint dynamics with Contact) Multi-Agent Soccer Environment [16] (MuJoCo is a physics engine designed for simulating complex, realistic motion and interactions in robotic and biomechanical systems). Recently, models have been proposed to train end-to-end robot soccer policies with fully onboard computation and sensing via egocentric RGB vision [37]. These advancements highlight the need for bipedal robot models to simulate sophisticated physical movements accurately. Future developments are expected to focus on enhancing real-to-sim techniques for bipedal robots, leveraging real-world pose data to improve the realism and effectiveness of these simulations. This will be important for advancing the field of sports analytics and achieving more accurate and reliable simulations.

Feedback and Enhancement

The significance of feedback and enhancement, which has not been extensively covered in this book, lies in their ability to enhance the effectiveness of sports training and performance analysis. Research on visualization techniques is crucial for delivering analytical results in a comprehensible format to users. For instance, papers in soccer [7], basketball [42], and team sports broadly [36], have extensively discussed various visualization methods that help coaches and analysts interpret complex sports data more effectively. Additionally, there is a growing body of research on feedback and training methods using VR (and more broadly, XR) technologies [10, 26]. These studies emphasize how immersive environments and real-time feedback can improve athletes' training experiences and outcomes.

Following the feedback process, the next critical step in team sports is what we refer to as "Enhancement". This involves a comprehensive approach to improving player performance and team strategies based on the insights gained from feedback, which can be employed in upcoming matches to exploit the opponent's weaknesses and enhance the team's strengths [15, 25]. Coaches and analysts must collaborate to design targeted training sessions that address the identified weaknesses and reinforce successful tactics. Additionally, the enhancement process may involve mental conditioning, recovery protocols, and the integration of new techniques and technologies to ensure continuous improvement (e.g., [27]). By systematically implementing these enhancement strategies, teams can effectively translate feedback into actionable improvements, ultimately leading to better performance in future games. While there is potential for this process to be automated through coaching AI in the future, it is currently believed that human communication skills and reliability surpass those of machines. Even with the introduction of AI, a system for human oversight would be necessary to ensure the effectiveness and trustworthiness of the enhancement process.

Advancements in data acquisition, modeling, simulation, feedback, and enhancement are driving the integration of real-world data and digital modeling in sports. These innovations are expected to transform how athletes train, how strategies are developed, and how performances are enhanced.

5.2.3 Long Future Prospects and Benefits of Digital Modeling in Sports

The integration of real-world data with digital modeling has the potential to transform sports analytics, offering unprecedented insights and advancements across various domains. From enhanced performance monitoring and training programs to strategic decision-making and advanced recruitment strategies, the applications of learning-based sports analytics are vast and multifaceted. This chapter explores the innovative ways in which these technologies are reshaping the landscape of team sports, providing detailed analysis and new opportunities for coaches, players, and fans alike. By leveraging cutting-edge machine learning techniques, we can achieve new levels of understanding and efficiency, ultimately driving the evolution of sports analytics to new heights.

- **Enhanced performance monitoring**: The integration of real-world data with digital models will enable continuous and detailed monitoring of athlete performance. Currently, this monitoring primarily relies on biometric, GPS, and Inertial Measurement Unit (IMU) data (e.g., [32]). However, there is an increasing expectation that performance-based monitoring from images will also be able to track skill levels and identify injury risk factors. This advancement will allow for near real-time feedback and adjustments, helping athletes refine their skills and prevent injuries. Additionally, continuous monitoring can extend to tracking recovery processes and predicting peak conditions, providing insights into minute aspects of performance and health that were previously undetectable.
- **Refining training programs**: Training programs are structured plans designed to improve an athlete's performance through specific exercises, drills, and recovery protocols. Currently, these programs are developed by a range of professionals, including tactical and performance coaches, strength and conditioning coaches, and trainers. Each of these experts contributes their specialized knowledge to create comprehensive training programs. If we can digitalize their data as input, machine learning models can flexibly simulate different training scenarios and analyze their outcomes. This allows for the design of personalized training programs that maximize an athlete's potential. These programs can be dynamically adjusted based on near real-time performance data.
- **Strategic decision-making in game preparation, execution, and review**: Digital modeling can simulate game scenarios and analyze the effectiveness of different strategies, providing invaluable insights for coaches and analysts. Before the game, it aids in scouting opponents and selecting optimal lineups for one's own team. During the game, it helps in recognizing the opponent's tactics, predicting the flow of the match, and making timely decisions on substitutions and tactical adjustments. After the game, it allows for detailed analysis and validation of player movements and tactical decisions. Looking ahead, the ability to run complex simulations in near real-time could enable adaptive strategy adjustments during games, giving teams a significant competitive edge.

- **Advanced recruitment strategies**: Traditionally, the recruitment of promising young athletes or competitive players from other leagues and teams has relied on video footage, static statistics related to ball handling and athletic performance, and subjective assessments by scouts, agents, and management. However, with learning-based sports analytics, it is now possible to conduct deeper evaluations of players' in-game actions, such as effective off-ball offensive movements and coordinated defensive plays. These insights allow for more comprehensive assessments of a player's overall contribution and potential for growth. Consequently, teams can better identify players who fit specific roles and strategies, leading to more targeted and effective recruitment decisions that align with the team's long-term goals.
- **Learning-based refereeing**: The future of learning-based (or AI) refereeing in team sports is promising, with advancements focusing on two main aspects: motion detection and evaluation. Motion detection typically involves estimating the positions and postures of players, such as the location of their joints and their position on the field, often using non-contact methods like cameras. In the evaluation phase, where detected motions are assessed, AI referees are most effective in scenarios with clearly defined rules and measurable parameters. Well-known applications include offside detection in soccer (e.g., [39]). However, more complex fouls or violations, such as contact fouls in soccer and basketball, pose significant challenges. In motion detection, the crowded nature of such scenes often leads to occlusion issues, complicating motion estimation. In addition, these situations often involve subjective judgment and player tactics that make accurate detection difficult. For instance, players might exaggerate their movements to simulate a foul, complicating the machine learning task. Although AI can assist by detecting initial contact, fully automating foul detection remains challenging. Research in this area is active, with new models being developed to explain foul decisions using large language models [17]. In the near future, AI systems may be integrated to support human referees, aiding in training and providing supplementary foul assessments.
- **Fan engagement and experience**: In the field of fan engagement, advancements in sports analytics can significantly enhance how fans interact with broadcast footage, websites, and other media. Beyond merely highlighting scoring plays, these technologies can quantitatively explain the decision-making processes of players, making it easier for fans to understand the rationale behind key actions. This includes extracting and evaluating the performances of favorite players, even when they are away from the ball. Such detailed and quantitative analysis can promote new levels of fan engagement by providing deeper insights into the game. In particular, this approach offers unbiased, quantitative commentary on increasingly sophisticated tactics, making it easier for beginners to understand complex strategies. Similarly, in the context of in-game betting, establishing real-time evaluation systems can enhance the accuracy of predictions and enrich the overall betting experience, providing fans with more reliable and engaging interactions.
- **Democratizing sports analytics via machine learning**: Currently, data acquisition systems are predominantly used in well-funded professional sports leagues,

such as European soccer and basketball in the US. The high costs associated with obtaining this data often result in it being restricted by companies and leagues, limiting public access and restricting the democratization of sports analytics. Alternatively, automated data acquisition through video processing from a limited number of cameras can reduce these costs. If this technology becomes widely accessible, it has the potential to democratize sports analytics. This would be particularly beneficial for amateur and youth athletes, enabling them to access advanced analytics and improve their performance. Such democratization would allow everyone to enjoy the benefits of sophisticated sports analytics, developing a more inclusive and equitable sports environment. Further discussion on this topic will be provided in the next section.

In summary, the integration of real-world data and digital modeling will change sports analytics by providing advanced tools for performance monitoring, refining training, strategic decision-making, and injury prevention. These technologies offer significant benefits, from enhancing fan engagement and democratizing sports analytics to supporting advanced recruitment strategies and learning-based refereeing. As these innovations continue to develop, they will play an increasingly vital role in shaping the future of sports and developing a more inclusive and equitable environment. The next section will focus on the practical deployment and real-world implementation of these advancements in sports analytics, examining how they can be effectively utilized in various settings to maximize their impact.

5.3 Practical Deployment of Learning-Based Analytics

The practical deployment and real-world implementation of sports analytics involve more than just the technical rollout of systems; they encompass the integration of these technologies into everyday practices and workflows. This process includes not only the technical aspects of installing and configuring the systems but also ensuring that they are effectively utilized by end-users such as coaches, players, analysts, and other stakeholders. The significance of practical deployment and real-world implementation lies in its ability to transform theoretical models and analytical tools into actionable insights and strategies that can enhance performance, improve decision-making, and provide a competitive edge in sports.

5.3.1 Data Acquisition

When filming with a single camera for sports, which requires coverage of a large field such as soccer and rugby, it is necessary to move the camera and zoom in on the ball and surrounding players, similar to broadcast footage. This scenario requires the use of field registration techniques, as explained in Chap. 2. If the camera can be fixed, the need for field registration is minimized, making the estimation problem much simpler and less time-consuming, even feasible to do manually. Despite this advantage, a fixed camera setup results in smaller images of players and the ball, which complicates tracking and re-identification (Re-ID). Therefore, a practical solution may involve using two fixed cameras, balancing the trade-off between manual effort and the accuracy of computer vision techniques.

In many sports, the ball can often be obscured by players (though soccer tends to have better visibility), but if event data can be obtained, linking it to the ball handler's ID often negates the need for precise ball position data. Realistically, tracking all players and annotating events require human correction of machine learning predictions. Thus, a user-friendly interface that reduces manual labor is essential. For example, an event annotation tool for soccer[1] is publicly available, and that interface for basketball is illustrated in [41]. For pose data, manual correction is almost cost-prohibitive, thus methods that can analyze using incomplete estimation data are necessary.

Using GPS sensors to estimate player positions can help avoid occlusion problems, but it comes with potential issues of drift and operational costs related to attachment and removal. Additionally, other sensors such as IMUs and heart rate monitors can provide valuable information about player movements and physiological states. However, deploying these sensors requires careful planning to avoid interfering with player performance. Integrating sensor data with analytical platforms involves setting up efficient data pipelines, ensuring data accuracy, and addressing issues related to sensor calibration and synchronization. Proper handling of these aspects is crucial to effectively leverage sensor data for enhancing player performance analysis and overall game strategy.

5.3.2 Application Examples in Learning-Based Sports Analytics

After acquiring data, machine learning-based analytics are applied according to various on-field objectives. The techniques employed differ based on user goals and are thus diverse. Currently, developing a universal solution to address all these needs is challenging. Therefore, This section discusses the technically feasible applications of these technologies in the same order as presented in Sect. 5.2.2. Each example will illustrate how current machine learning methods can be utilized to enhance different aspects of sports performance and strategy, demonstrating their practical deployment in real-world scenarios.

- **Enhanced performance monitoring**: Currently, performance monitoring in sports primarily relies on metrics such as total distance covered during a match and the number of sprints, which can be also calculated using GPS. Additionally, mechan-

---

[1] https://github.com/SilvioGiancola/SoccerNetv2-DevKit/tree/main/Annotation.

ical and physiological loads can be assessed using IMUs and heart rate monitors, while event data can provide information on the number of passes made. However, with the availability of positional data for all players captured by cameras, it becomes possible to calculate more detailed metrics. For instance, the speed of sprints and the number of passes made in specific tactical situations can be accurately measured. This level of detail allows for a more comprehensive understanding of player performance and tactical execution, enabling coaches and analysts to make more informed decisions to optimize team strategies and individual player development.

- **Strategic decision-making in game analysis**: Currently, the technology available for capturing accurate tracking data in real-time is incomplete for live game analysis, thus limiting the primary use of analytics to pre-game preparations and post-game reviews. In pre-game analysis, accumulated data can be leveraged to understand the opponent's tactics and key players beyond basic statistics. It also allows for evaluating the recent tactical success rates and performance levels of one's own players, aiding in strategic decision-making for the next game, such as selecting tactics and lineups. Post-game reviews focus on verifying the effectiveness of chosen strategies and lineups and preparing for future matches. This process does not necessarily require extensive data from multiple games, though having more data can enhance predictions and digital modeling, thereby broadening the scope and depth of analysis.

- **Advanced recruitment strategies**: Recruiting promising young athletes or competitive players from other leagues and teams, along with providing valuable insights to agents and executives, requires a vast database facilitated by extensive networking. Traditionally, this recruitment process has relied on video footage, static statistics related to ball handling and athletic performance, and subjective assessments. While commercial services exist that share such data internationally, they often lack comprehensive tracking data, especially for off-ball movements. To implement learning-based analytics that includes these off-ball actions, standardized methods for collecting and sharing tracking data are essential. This presents a significant challenge, as current practices are not yet equipped to handle such extensive data acquisition and standardization. To overcome this, collaboration between leagues, teams, and technology providers is necessary to develop and adopt unified tracking systems, ensuring consistent data quality and accessibility across different regions and competitions.

- **Referee assistance systems**: In evaluating player movements from the captured data, referee assistance systems are most effective in scenarios where position data can be accurately measured, and the movements are clearly defined by human standards. Known applications include gymnastics (e.g., [2]) and offside detection (e.g., [40]) in soccer, and more advanced systems such as automated line judgment in volleyball (e.g., [29]) and tennis (e.g., [9]), and automated strike zone calls in baseball (e.g., [23]) are already in use. The feasibility of implementing referee assistance systems hinges on both the current technological capabilities and the practicality of their integration into the existing framework. From a technological perspective, research is ongoing in various sports, including swimming dives [43], figure skating jumps [35], and race walk faults [34] detection. While the precision required for practical deployment varies, achieving the necessary accuracy for competition use will demand extensive data acquisition and refinement of AI models, necessitating larger teams and resources. Although technically feasible, as demonstrated in gymnastics, integrating these systems into official events presents challenges such as cost, incorporation into existing rules, the interaction between human referees and AI, and ensuring the systems do not detract from the excitement of the game. Thus, implementing these systems requires substantial effort and collaboration among stakeholders. Moreover, the accuracy of operational referee assistance systems is expected to match or exceed human referees, given their multi-layered validation process. Human referees bring extensive experience but are limited by their biological constraints, while AI systems can process data with high spatiotemporal resolution and operate without time and location restrictions, making them potentially more effective in many scenarios.

- **Fan engagement and experience**: To enhance the fan experience or provide valuable information for betting, we need to move beyond the current methods centered around video footage, static statistics, and heat maps. Learning-based analytics that evaluates all players in real-time is essential for this advancement. While it is technically feasible to perform field registration, tracking, and Re-ID in real time on broadcast footage at the cost of some accuracy, the main challenge lies in determining what information to provide to viewers and bettors. For those receiving information in real-time, too much data can overwhelm and detract from the experience. Therefore, it is important to carefully curate and minimize the information presented to maintain an engaging and informative user experience. Further considerations include designing user interfaces that highlight key insights and ensuring the information is actionable and relevant, thereby enriching the overall engagement and making the viewing or betting experience more enjoyable and insightful.

Despite the current capabilities of technology, there are numerous opportunities for deployment. In well-funded professional sports leagues, data companies possess vast amounts of data, facilitating machine learning-based research. However, the increasing cost and strict management of broadcasting rights, along with the large number of stakeholders and organizational rigidity, often hinder new initiatives. On the other hand, amateur leagues and university teams, though limited in data volume (typically covering only a few games), can benefit from more flexible and rapid deployment. Open innovation can promote frequent information transfers between amateur and professional levels, accelerating the adoption of learning-based analytics across all tiers of sports. This exchange of knowledge and techniques between amateurs and professionals will enhance the overall landscape of sports analytics, making it more accessible and widespread.

5.3.3 Key Principles for Advancing Learning-Based Sports Analytics

To advance learning-based sports analytics effectively, it is essential to support open innovation that involves those who collect data, those who analyze it, and those who utilize it. This approach relies on establishing several key principles, such as trustworthiness, fairness, and transparency. These principles ensure the reliable and ethical development and deployment of analytics systems, promoting wider acceptance and use. By integrating these concepts, we can enhance the credibility and utility of sports analytics, ultimately benefiting all stakeholders involved.

- **Trustworthiness**: Trustworthiness refers to the inherent qualities and characteristics that make a system or organization reliable and dependable. In the context of sports analytics, systems must demonstrate consistency, accuracy, and integrity in their operations and outputs. By ensuring that data acquisition, processing, and analysis are conducted with the highest standards of reliability, users can develop confidence in the analytics provided. Trustworthiness is the foundation on which trust is built, making it essential for the adoption and acceptance of advanced sports analytics systems.
- **Fairness**: Fairness in sports analytics involves ensuring that the algorithms and models reduce biases and that they provide equitable opportunities for all athletes and teams. This includes addressing potential differences in data representation and algorithmic decisions that could unfairly disadvantage certain groups. For example, in the case of AI referees, it is important that these systems are designed and trained to avoid any biases that might result in inconsistent or unfair rulings, particularly against underrepresented teams or players. By prioritizing fairness, we can promote a level playing field and ensure that analytics contribute positively to the development and success of all participants, regardless of their background or circumstances.
- **Transparency**: Transparency is critical for building trust and ensuring accountability in sports analytics. This involves openly sharing the methodologies, data sources, and decision-making processes used in analytics systems. By making these aspects accessible and understandable to users, analysts, and stakeholders, we can clarify the technology and promote greater acceptance. Transparency also allows for independent verification and validation, further enhancing the credibility and reliability of the analytics provided.

By focusing on these principles-trustworthiness, fairness, transparency, and collaboration-we can create a robust framework for the ethical and effective advancement of learning-based sports analytics. This foundation will support the growth of the field, encourage widespread adoption, and ultimately lead to better performance insights and outcomes in sports. Based on these principles, the next section will introduce specific proposals for shaping future ecosystems in sports analytics.

5.4 Shaping Future Ecosystems

To fully realize the future applications of sports analytics introduced in Sect. 5.2 and to enhance the current implementations discussed in Sect. 5.3, it is essential to shape supportive ecosystems that involves those who collect data, those who analyze it, and those who utilize it. These ecosystems are crucial for promoting innovation, ensuring standardization, and promoting ethical use. By building collaborative research networks, promoting open source initiatives, and standardizing data formats within sports analytics, we can create a robust foundation for future advancements. For example, PySport for various python libraries in many sports, SportsLabKit[2] and TrackLab,[3] which processes sports videos, and OpenSTARLab[4] and Kloppy, which plans to further enhance its library for team sports advanced analysis in the future, can be mentioned. Additionally, addressing cross-disciplinary innovations, sustainability, educational programs, and public policy outside of sports analytics will not only drive technological progress but also increase the user base and acceptance of these technologies. These efforts are necessary to build a sustainable, innovative, and ethically sound ecosystem that supports the long-term growth and impact of sports analytics.

5.4.1 Building the Infrastructure of Sports Analytics

The development of a robust ecosystem within sports analytics is essential to achieving long-term success and sustainability in the field. Such an ecosystem facilitates collaboration, innovation, and the seamless integration of various technologies and methodologies. By focusing on collaborative research networks, open data initiatives, and the standardization of data formats, we can create an environment that leads to significant advances in sports analytics.

- **Open source initiatives**: Promoting the sharing of sports data and code for research and development is essential for accelerating progress in sports analytics. Open source initiatives encourage transparency and accessibility, allowing researchers and developers to access a wealth of information that can be used to train models, validate theories, and develop new applications. By making data widely available, these initiatives reduce barriers to entry, enabling more individuals and organizations to contribute to the field. Furthermore, open data cultivates a culture of collaboration and innovation, as researchers can build on each other's work, leading to more rapid and impactful advancements. To further enhance these efforts, organizing competitions and open lectures can play a significant role in promoting understanding and engagement. Competitions can challenge participants to

---

[2] https://github.com/AtomScott/SportsLabKit.

[3] https://github.com/TrackingLaboratory/tracklab.

[4] https://github.com/open-starlab.

develop innovative solutions using open data and code, while open lectures provide educational opportunities to learn from experts in the field. These activities not only drive interest and participation but also help disseminate knowledge and best practices, furthering the goals of open source initiatives.

- **Standardization of data formats**: Establishing common protocols for data acquisition and analysis in sports is vital for ensuring consistency and interoperability across different systems and platforms. Standardization simplifies the integration of various data sources, making it easier to combine and analyze information from different sensors, devices, and applications. This uniformity enhances the reliability and comparability of analytical results, enabling more accurate and actionable insights. For instance, tracking and pose estimation benefit from common formats such as MOT (multi-object tracking) and COCO (common objects in context), which have facilitated concentrated research and algorithm development. In the sports domain, initiatives like PySport[5] provide diverse Python code for sports analytics, but there are few efforts to create common formats for different datasets. Kloppy[6] is a notable exception, as it aims to standardize event and tracking data from various soccer data providers into a common format. Similarly, the Events in Invasion Games Dataset [5] attempts to represent event data for handball and soccer using a unified format. By adopting standardized data formats, the sports analytics community can streamline the development and deployment of new technologies, reduce redundancy, and increase efficiency.

- **Collaborative research networks**: Building networks between academic institutions, sports organizations, and tech companies are crucial for promoting innovation and advancing research in sports analytics. These networks enable the sharing of knowledge, resources, and expertise, leading to the development of more sophisticated analytical tools and techniques. Collaboration also helps bridge the gap between theoretical research and practical application, ensuring that innovations are grounded in real-world needs and challenges. Academic conferences play a vital role in this ecosystem by providing a platform for presenting cutting-edge research, exchanging ideas, and forming partnerships. However, the increase of conferences and workshops has made it challenging to consolidate efforts and maintain focus. Therefore, establishing a larger community that integrates the efforts of academic conferences with corporate partnerships is essential. This integrated approach can help streamline efforts, reduce redundancy, and enhance the impact of collaborative research networks. By facilitating these partnerships, such networks can drive the evolution of sports analytics, enabling stakeholders to leverage the collective intelligence and capabilities of the community effectively.

---

[5] https://opensource.pysport.org/.

[6] https://kloppy.pysport.org/.

These foundational elements within sports analytics are essential for creating a robust ecosystem that supports continuous innovation and application. By promoting collaboration, promoting open data, and standardizing data formats, we can ensure that sports analytics remains at the forefront of technological advancement, driving improvements in athletic performance, fan engagement, and overall sports management.

5.4.2 *Extending the Reach of Sports Analytics*

To maximize the impact and reach of sports analytics, it is crucial to consider broader contexts beyond the immediate domain. This involves promoting cross-disciplinary innovations, developing educational programs, and shaping public policy and regulation. These initiatives are essential for creating a holistic ecosystem that supports the integration and ethical application of sports analytics across various sectors.

- **Cross-disciplinary innovations**: Encouraging collaboration between fields such as robotics, artificial intelligence, and sports science is essential for driving innovation in sports analytics. By leveraging the expertise and methodologies from different disciplines, we can develop more comprehensive and effective solutions. For example, the success of digital modeling in sports analytics fundamentally depends on the integration of robotics and machine learning. Although neither discipline can fully replicate human abilities at present, both are capable of approximating human cognition and movement to a substantial degree. It is important to approach these fields not as isolated domains but as interconnected areas of research. Robotics, which emphasizes physical interaction with the environment, and machine learning, which excels in processing and predicting complex patterns, together play a vital role in narrowing the gap between human intuition and artificial replication. The continued development of these fields, in conjunction with one another, will be essential for advancing our understanding and modeling of human performance in sports.
- **Educational programs**: Developing curricula to train the next generation of sports data scientists and analysts is essential for ensuring continued growth and innovation in the field. Educational programs should focus on providing students with a strong foundation in data science, machine learning, and sports analytics, as well as practical experience with real-world data and tools. By equipping students with the necessary skills and knowledge, we can cultivate a workforce capable of advancing sports analytics and applying it to a wide range of applications. This investment in education will help sustain the field's momentum and ensure a steady supply of talented professionals.
- **Ethical governance of AI in sports decision-making**: Shaping public policy and regulation for the ethical use of AI and data in sports is important, particularly in ensuring fairness and transparency, which are of paramount importance in AI-based decision-making processes, such as AI refereeing. As AI and

data increasingly influence human decision-making in sports, policymakers must collaborate with researchers, sports organizations, and technology companies to establish guidelines that promote responsible use. This includes setting standards for data protection, ensuring transparency and accountability in AI algorithms, and addressing potential biases in data acquisition and analysis. Effective regulation will not only safeguard privacy but also build trust in sports analytics, ensuring that its benefits are achieved ethically and with social responsibility.

By focusing on these areas outside of sports analytics, we can create a supportive ecosystem that enhances the integration, application, and ethical use of sports analytics. These efforts will drive innovation, promote sustainability, and ensure the field's long-term growth and impact across various sectors.

5.5   Summary

This final chapter proposes a hypothesis about the future direction of learning-based sports analytics, emphasizing the integration of real-world data and digital modeling. This approach addresses challenges such as automating data acquisition, bridging domain gaps, and implementing learned models. The integration aims to enhance the prediction, evaluation, and optimization of play. The chapter also highlights the long-term benefits of digital modeling, including enhanced performance monitoring and training, strategic decision-making, advanced recruitment, and fan engagement. Practical deployment challenges are discussed, along with the need for open innovation and key principles like trustworthiness, fairness, and transparency. Shaping future ecosystems involves building infrastructure, fostering cross-disciplinary innovations, and addressing the ethical governance of AI in sports decision-making.

References

1. Ai, S., Na, J., De Silva, V., Caine, M.: A novel methodology for automating spatio-temporal data classification in basketball using active learning. In: 2021 IEEE 2nd International Conference on Pattern Recognition and Machine Learning (PRML), pp. 39–45. IEEE (2021)
2. Allen, E., Fenton, A., Parry, K.D.: Computerised gymnastics judging scoring system implementation–an exploration of stakeholders' perceptions. Sci. Gymnast. J. **13**(3), 357–370 (2021)
3. Anderson, A., Rosen, J., Rust, J., Wong, K.-P.: Disequilibrium play in tennis. J. Polit. Econ. (2024)
4. Anzer, G., Bauer, P., Brefeld, U., Faßmeyer, D.: Detection of tactical patterns using semi-supervised graph neural networks. In: MIT Sloan Sports Analytics Conference, vol. 16, pp. 1–3 (2022)

5. Biermann, H., Theiner, J., Bassek, M., Raabe, D., Memmert, D., Ewerth, R.: A unified taxonomy and multimodal dataset for events in invasion games. In: Proceedings of the 4th International Workshop on Multimedia Content Analysis in Sports, pp. 1–10 (2021)
6. Cabado, B., Cioppa, A., Giancola, S., Villa, A., Guijarro-Berdinas, B., Padrón, E.J., Ghanem, B., Van Droogenbroeck, M.: Beyond the premier: assessing action spotting transfer capability across diverse domains. In: Proceedings of the IEEE/CVF Conference on Computer Vision and Pattern Recognition, pp. 3386–3398 (2024)
7. Cao, A., Xie, X., Zhou, M., Zhang, H., Mingliang, X., Yingcai, W.: Action-evaluator: a visualization approach for player action evaluation in soccer. IEEE Trans. Visual Comput. Graph. **30**(1), 880–890 (2024)
8. Carion, N., Massa, F., Synnaeve, G., Usunier, N., Kirillov, A., Zagoruyko, S.: End-to-end object detection with transformers. In: Proceedings of the 16th European Conference on Computer Vision, pp. 213–229. Springer (2020)
9. Collins, H., Evans, R.: You cannot be serious! public understanding of technology with special reference to "hawk-eye". Public Underst. Sci. **17**(3), 283–308 (2008)
10. Cossich, V.R.A., Carlgren, D., Holash, R.J., Katz, L.: Technological breakthroughs in sport: current practice and future potential of artificial intelligence, virtual reality, augmented reality, and modern data visualization in performance analysis. Appl. Sci. **13**(23), 12965 (2023)
11. Fassmeyer, D., Fassmeyer, P., Brefeld, U.: Semi-supervised generative models for multiagent trajectories. Adv. Neural. Inf. Process. Syst. **35**, 37267–37281 (2022)
12. Fujii, K., Takeishi, N., Tsutsui, K., Fujioka, E., Nishiumi, N., Tanaka, R., Fukushiro, M., Ide, K., Kohno, H., Yoda, K., Takahashi, S., Hiryu, S., Kawahara, Y.: Learning interaction rules from multi-animal trajectories via augmented behavioral models. In: Advances in Neural Information Processing Systems, vol. 34, pp. 11108–11122 (2021)
13. Fuller, A., Fan, Z., Day, C., Barlow, C.: Digital twin: enabling technologies, challenges and open research. IEEE Access **8**, 108952–108971 (2020)
14. Giancola, S., Cioppa, A., Georgieva, J., Billingham, J., Serner, A., Peek, K., Ghanem, B., Van Droogenbroeck, M.: Towards active learning for action spotting in association football videos. In: Proceedings of the IEEE/CVF Conference on Computer Vision and Pattern Recognition, pp. 5098–5108 (2023)
15. Gilbert, W.: Coaching better every season: a year-round system for athlete development and program success. Hum. Kinet. (2016)
16. Haarnoja, T., Moran, B., Lever, G., Huang, S.H., Tirumala, D., Humplik, J., Wulfmeier, M., Tunyasuvunakool, S., Siegel, N.Y., Hafner, R., et al.: Learning agile soccer skills for a bipedal robot with deep reinforcement learning. Sci. Robot. **9**(89), eadi8022 (2024)
17. Held, J., Itani, H., Cioppa, A., Giancola, S., Ghanem, B., Van Droogenbroeck, M.: X-vars: introducing explainability in football refereeing with multi-modal large language models. In: Proceedings of the IEEE/CVF Conference on Computer Vision and Pattern Recognition, pp. 3267–3279 (2024)
18. Hong, J., Zhang, H., Gharbi, M., Fisher, M., Fatahalian, K.: Spotting temporally precise, fine-grained events in video. In: European Conference on Computer Vision, pp. 33–51. Springer (2022)
19. Kaplan, J., McCandlish, S., Henighan, T., Brown, T.B., Chess, B., Child, R., Gray, S., Radford, A., Wu, J., Amodei, D.: Scaling laws for neural language models (2020). arXiv:2001.08361
20. Kolter, J., Abbeel, P., Ng, A.: Hierarchical apprenticeship learning with application to quadruped locomotion. In: Advances in Neural Information Processing Systems, vol. 20 (2007)
21. Kurach, K., Raichuk, A., Stańczyk, P., Zając, M., Bachem, O., Espeholt, L., Riquelme, C., Vincent, D., Michalski, M., Bousquet, O., et al.: Google research football: a novel reinforcement learning environment. In: Proceedings of the AAAI Conference on Artificial Intelligence, vol. 34, pp. 4501–4510 (2020)
22. Le, H.M., Yue, Y., Carr, P., Lucey, P.: Coordinated multi-agent imitation learning. In: Proceedings of the 34th International Conference on Machine Learning, vol. 70, pp. 1995–2003 (2017). (JMLR. org, 2017)

23. Lee, K., Han, K., Ko, J.: Analyzing the impact of the automatic ball-strike system in professional baseball: a case study on KBO league data (2024). arXiv:2407.15779
24. Li, C., Wang, T., Chengjie, W., Zhao, Q., Yang, J., Zhang, C.: Celebrating diversity in shared multi-agent reinforcement learning. Adv. Neural. Inf. Process. Syst. **34**, 3991–4002 (2021)
25. Martens, R., Vealey, R.S.: Successful Coaching. Hum. Kinet. (2024)
26. Miles, H.C., Pop, S.R., Watt, S.J., Lawrence, G.P., John, N.W.: A review of virtual environments for training in ball sports. Comput. Graph. **36**(6), 714–726 (2012)
27. Russell, M., West, D.J., Harper, L.D., Cook, C.J., Kilduff, L.P.: Half-time strategies to enhance second-half performance in team-sports players: a review and recommendations. Sports Med. **45**, 353–364 (2015)
28. Rusu, A.A., Večerík, M., Rothörl, T., Heess, N., Pascanu, R., Hadsell, R.: Sim-to-real robot learning from pixels with progressive nets. In: Conference on Robot Learning, pp. 262–270. PMLR (2017)
29. Sarvestan, J., Khalafi, M.: Smart line judgement system: a novel technology in vol-leyball arbitration. Insight Mater. Sci. **2**(1), 1–7 (2019)
30. Schaal, S.: Learning from demonstration. Adv. Neural. Inf. Process. Syst. **9**, 1040–1046 (1996)
31. Scott, A., Uchida, I., Ding, N., Umemoto, R., Bunker, R., Kobayashi, R., Koyama, T., Onishi, M., Kameda, Y., Fujii, K.: Teamtrack: a dataset for multi-sport multi-object tracking in full-pitch videos. In: Proceedings of the IEEE/CVF Conference on Computer Vision and Pattern Recognition, pp. 3357–3366 (2024)
32. Seçkin, A.Ç., Ateş, B., Seçkin, M.: Review on wearable technology in sports: concepts, challenges and opportunities. Appl. Sci. **13**(18), 10399 (2023)
33. Shi, J., Wan, J., Yan, H., Suo, H.: A survey of cyber-physical systems. In: 2011 International Conference on Wireless Communications and Signal Processing (WCSP), pp. 1–6. IEEE (2011)
34. Suzuki, T., Takeda, K., Fujii, K.: Automatic detection of faults in simulated race walking from a fixed smartphone camera. Int. J. Comput. Sci. Sport **23**(1), 22–36 (2024)
35. Tanaka, R., Suzuki, T., Fujii, K.: 3d pose-based temporal action segmentation for figure skating: a fine-grained and jump procedure-aware annotation approach. In: Proceedings of the 7th International Workshop on Multimedia Content Analysis in Sports (2024)
36. Thornton, H.R., Delaney, J.A., Duthie, G.M., Dascombe, B.J.: Developing athlete monitoring systems in team sports: data analysis and visualization. Int. J. Sports Phys. Perform. **14**(6), 698–705 (2019)
37. Tirumala, D., Wulfmeier, M., Moran, B., Huang, S., Humplik, J., Lever, G., Haarnoja, T., Hasenclever, L., Byravan, A., Batchelor, N., et al.: Learning robot soccer from egocentric vision with deep reinforcement learning (2024). arXiv:2405.02425
38. Tsutsui, K., Takeda, K., Fujii, K.: Synergizing deep reinforcement learning and biological pursuit behavioral rule for robust and interpretable navigation. In: 1st Workshop on the Synergy of Scientific and Machine Learning Modeling in International Conference on Machine Learning (2023)
39. Uchida, I., Scott, A., Shishido, H., Kameda, Y.: Automated offside detection by Spatio-Temporal analysis of football videos. In: Proceedings of the 4th International Workshop on Multimedia Content Analysis in Sports. Association for Computing Machinery, pp. 17–24 (2021)
40. Viswanathan, S., Shankar, R., Jackson, J., Binns, R.: Spectators of ai: football fans vs. the semi-automated offside technology at the 2022 fifa world cup. In: Extended Abstracts of the 2023 CHI Conference on Human Factors in Computing Systems, pp. 1–6 (2023)
41. Wu, T., He, R., Wu, G., Wang, L.: Sportshhi: a dataset for human-human interaction detection in sports videos. In: Proceedings of the IEEE/CVF Conference on Computer Vision and Pattern Recognition, pp. 18537–18546 (2024)
42. Yihong, W., Deng, D., Xie, X., He, M., Jie, X., Zhang, H., Zhang, H., Yingcai, W.: Obtracker: visual analytics of off-ball movements in basketball. IEEE Trans. Visual Comput. Graph. **29**(1), 929–939 (2023)
43. Xu, J., Rao, Y., Yu, X., Chen, G., Zhou, J., Lu, J.: Finediving: a fine-grained dataset for procedure-aware action quality assessment. In: Proceedings of the IEEE/CVF Conference on Computer Vision and Pattern Recognition, pp. 2949–2958 (2022)

44. Zheng, S., Yue, Y., Hobbs, J.: Generating long-term trajectories using deep hierarchical networks. In: Advances in Neural Information Processing Systems, vol. 29, pp. 1543–1551 (2016)
45. Ziyi, Z., Takeda, K., Fujii, K.: Cooperative play classification in team sports via semi-supervised learning. Int. J. Comput. Sci. Sport **21**(1), 111–121 (2022)

Glossary

Data Collection The process of collecting data from various sources such as video footage, wearable sensors, and GPS tracking devices in sports analytics.

Data Preprocessing The process of cleaning, transforming, and organizing raw data into a usable format.

Feature Engineering The process of using domain knowledge to extract meaningful features from raw data that enhance the performance of machine learning models.

Model Development The process of selecting and building appropriate machine learning and statistical models tailored to specific tasks in sports analytics.

Model Evaluation The process of assessing the performance of models through various metrics to ensure reliability and accuracy in predictions within sports analytics.

Data Visualization The representation of data in graphical formats such as graphs, tables, maps, and charts.

Strategies Comprehensive plans designed to achieve long-term objectives in sports.

Tactics Specific actions or sequences of actions that teams employ during a game to execute their strategy.

Techniques The fundamental skills required to execute tactics effectively in sports.

Forward Problem The process of generating outcomes from known causes or models.

Inverse Problem The process of deducing the underlying causes or strategies from observed outcomes.

Mathematical Models The use of mathematical expressions to represent real-world phenomena.

Rule-Based Models An approach where the underlying rules governing a system or behavior are explicitly programmed into a computer by humans.

Machine Learning Models Algorithms that automatically extract useful patterns or knowledge from input data.

Unsupervised Learning A type of machine learning that involves learning without using target variables.

© The Editor(s) (if applicable) and The Author(s) 2025   123
K. Fujii, *Machine Learning in Sports*,
SpringerBriefs in Computer Science, https://doi.org/10.1007/978-981-96-1445-5

Supervised Learning A type of machine learning that involves learning to align with target variables.

Reinforcement Learning A type of machine learning where agents learn to make decisions by interacting with their environment and receiving rewards or penalties.

Counterfactual Analysis A technique in sports analytics that explores hypothetical scenarios to understand the impact of different strategies and player movements on game outcomes.

Trajectory Prediction The process of predicting the future positions of players during a game using machine learning models, often applied in team sports scenarios.

Field Registration The alignment of captured video footage with a pre-defined playing field.

Camera Calibration The process of determining the parameters of a camera, enabling the transformation between 3D world coordinates and 2D image coordinates.

Tracking The process of detecting and following the movements of objects, such as players and the ball, across successive video frames.

Re-identification The process of recognizing the same player across different camera views or after occlusions.

Action Recognition and Detection The process of identifying specific actions or events in sports footage.

Pose Estimation The process of determining the precise positions of a player's body joints from video footage.

Index